WA

Handbook o

D1435120

SECOND EDITION

HANDBOOK OF POISONOUS AND INJURIOUS PLANTS

Lewis S. Nelson, M.D.
Richard D. Shih, M.D.
Michael J. Balick, Ph.D.

Foreword by Lewis R. Goldfrank, M.D.
Introduction by Andrew Weil, M.D.

THE NEW YORK
BOTANICAL GARDEN

Springer

Lewis S. Nelson, MD
New York University
  School of Medicine
New York City Poison
  Control Center
New York, NY 10016
USA

Richard D. Shih, MD
New Jersey Medical School
Newark, NJ 07103
Morristown Memorial Hospital
Morristown, NJ 07962
Emergency Medical Associates
Livingston, NJ 07039
USA

Michael J. Balick, PhD
Institute of Economic Botany
The New York Botanical Garden
Bronx, NY, 10458
USA

Library of Congress Control Number: 2005938815

ISBN-10: 0-387-31268-4
ISBN-13: 978-0387-31268-2

e-ISBN-10: 0-387-33817-9
e-ISBN-13: 978-0387-33817-0

Printed on acid-free paper.

springer.com

Contents

Foreword

This second edition of the *Handbook of Poisonous and Injurious Plants* is a remarkable improvement to a great book. The first edition, *The AMA Handbook of Poisonous and Injurious Plants*, offered a major advance in linking high-quality botany, pharmacognosy, fine graphics, and a limited amount of clinical medicine. The first edition's authors, Kenneth Lampe and Mary Ann McCann, established a standard for quality at the interface between botany and medicine. Their vision, the increasing societal use of herbal preparations, and the investigatory spirit of those working in Poison Control Centers and emergency departments have led to the dramatic intellectual, organizational, and photographic advances seen in this second edition.

In the 20 years that have passed since the publication of the first edition, both Poison Control Centers and emergency departments have dramatically expanded their roles in society with regard to intentional and unintentional poisoning caused by plant exposure. The staff of emergency departments and poison control centers have developed close working relationships that have had salutary effects on clinical care and risk assessment in our country and throughout the world.

The authors of this second edition represent a fusion of clinical and botanical worlds. Drs. Lewis Nelson and Richard Shih are both physicians educated in emergency medicine and medical toxicology, and Dr. Michael Balick is a botanist trained in the study of useful and harmful plants. Their collaborative efforts have created a handbook that meets the clinician's needs. This text has enhanced the previous edition's botanical rigor with that of pharmacognosy and clinical medicine. This second edition is created to assist the clinician in addressing the needs of a poisoned child or adult. The authors have created a rigorous approach that starts with the physician addressing the patient's signs and symptoms by symptom complexes. In a structured fashion, the text suggests common plants that might lead to the development of the symptom complex and describes the mechanisms of action of the implicated toxin, additional clinical manifestations, and specific therapeutics for each presentation. The photographs of frequently encountered and clinically important plants are elegantly presented to permit the clinician to assist in the evaluation of potential toxic plant ingestions.

The botanical descriptions of the several hundred cited potentially toxic and injurious plants are precise, detailed, and ideally suited for all of us (clinicians and nonclinicians) who venture into the outdoors. This book will obviously be a vital resource not only for the staff of emergency departments and Poison Control Centers but also for many generalists. The barriers between botanists and physicians are so effectively broken down that the unreasonable fears and ignorance about plant ingestions are substantially alleviated. The authors' efforts should be pleasurable reading for all who enjoy investigating the beauty and unknown characteristics of our natural environment.

This book is a demonstration of how successful a cooperative venture can be among rigorous thinkers and investigators from different intellectual domains. This text will expand our understanding of nature while permitting more expeditious and higher-quality clinical care.

Lewis R. Goldfrank, M.D.
Professor and Chair
Department of Emergency Medicine
New York University School of Medicine
Director, Emergency Medicine
Bellevue Hospital/NYU Hospitals/VA Medical Center
Medical Director, New York City Poison Control Center

Acknowledgments

Many people contributed to the production of this book. It is a revision of the wonderful work of Kenneth F. Lampe and Mary Ann McCann, originally published in 1985 by the American Medical Association (AMA) as *The AMA Handbook of Poisonous and Injurious Plants*. Several years ago, that organization solicited a revised version of the book, and when this revision was completed, we learned that it was no longer within the subject matter published by that organization. Given our interest in seeing this volume appear in print, we requested that it be released for publication elsewhere, and the AMA was kind enough to agree to the release. One significant difference in this second edition is that toxic mushrooms are not discussed by us. Since the original publication, many fine books on mushrooms and their toxicity and management have appeared, such as *Mushrooms: Poisons and Panaceas: A Handbook for Naturalists, Mycologists, and Physicians* (Denis R. Benjamin, W.H. Freeman & Company, 1995), *Hallucinogenic and Poisonous Mushrooms: Field Guide* (Gary P. Menser, Ronin Publishing, 1996), and *National Audubon Society Field Guide to North American Mushrooms* (Gary A. Lincoff, Knopf, 1981), and given their depth and scope, we decided not to include this topic in the second edition.

The subject matter in this book reflects a partnership between botany and medicine, and many specialists were consulted from each area. We are grateful to Willa Capraro and Tom Zanoni for their contributions to the botanical side of the manuscript, updating some of the nomenclature and taxonomy used in the book. Irina Adam and Rebekka Stone Profenno worked diligently to secure and organize the photographs and drawings that appear in the book. The United States National Herbarium at the National Museum of Natural History (NMNH) kindly provided elements of the newly accessioned photographic collection of the late Harvard Professor Richard A. Howard, noted international authority on the botany of toxic plants, in order that they be available for this book. We thank George F. Russell of the NMNH for collaboration in that endeavor. Elizabeth Pecchia produced manuscript copy of the original book, and it was through her patient and capable labors that we were able to work from a typewritten copy of the original text. Some of the plant descriptions were based on information from Steven Foster and Roger Caras's book *Venomous Animals & Poisonous Plants* (The Peterson Field Guide Series, Houghton Mifflin

Company, New York, 1994), *Hortus Third* (L.H. Bailey and E.Z. Bailey and The Staff of the Liberty Hyde Bailey Hortorium, Macmillian Publishing Co., New York, 1976), and *Manual of Vascular Plants of Northeastern United States and Adjacent Canada*, Second Edition (Henry A. Gleason and Arthur Cronquist, The New York Botanical Garden, Bronx, New York, 2004), and we are most grateful to the authors of these important works. In contemporary times, websites are also valuable scholarly resources, and in working on this book botanical data were gathered from W³TROPICOS of The Missouri Botanical Garden (*http://mobot.mobot.org/W3T/Search/vast.html*), IPNI—The International Plant Names Index (*http://www.ipni.org/index.html*), and The New York Botanical Garden Virtual Herbarium (*http://sciweb.nybg.org/science2/VirtualHerbarium. asp*), as well as from numerous other sites on individual plants or images as necessary.

We are grateful to all the photographers who provided material for this book, in particular, Steven Foster, Richard W. Lighty, Irina Adam, and the late Richard A. Howard, as well as the third author; this group collectively contributed the bulk of the photographs used in this book. Others who provided photographs include Scooter Cheatham, Peter Goltra, Hans-Wilhelm Gromping, Flor Henderson, Andrew Henderson, Fredi Kronenberg, George K. Linney, John Mickel, Michael Nee, Kevin Nixon, Thomas Schoepke, and Dennis Wm. Stevenson. As a collection, these images have greatly enhanced the Second Edition, making it much more user friendly. We are grateful for Bobbi Angell's wonderful botanical illustrations that make the glossary so much more understandable. We turned to the excellent bibliographic resources of The LuEsther T. Mertz Library of The New York Botanical Garden in the search for plates of specific plants that were otherwise not available from the photographers we queried, and are grateful to the entire staff, in particular, Stephen Sinon and Marie Long, for their patient assistance in our search for appropriate illustrations. We thank the Archives of The New York Botanical Garden for use of photographic images from its collections. The New York Botanical Garden's living collections were an important resource for illustrating this book and for understanding the plants we discuss herein, and we are grateful to Carlo Balistrieri, Margaret Falk, Francesca Coelho, Todd Forrest, Jolene Yukes, and Kim Tripp for their help and interest in this project. Dennis Wm. Stevenson was generous in providing information on cycad toxicity and images, and William Buck, Scott Mori, and Michael Nee were kind enough to provide their insight on some of our botanical questions. Richard Schnall and the staff of Rosedale Nurseries allowed us to wander in their nursery and photograph interesting cultivars. Paul Schulick and Tom Newmark of New Chapter, Inc., very kindly made their Costa Rican farm, Luna Nueva, available to us for photographing.

Lewis Nelson and Richard Shih acknowledge Lewis Goldfrank, M.D., who spurred our initial interest in the toxic properties of poisonous plants during our years of medical training under his tutelage. Dr. Goldfrank would regularly bring to our workplace examples of poisonous plants that he found in his garden and enlighten us on the clinical implications of such exposure. We express our deep appreciation to Oliver Hung, M.D., and Richard Hamilton, M.D., both of whom provided insight during both the formative and final stages of book production.

Michael Balick wishes to thank two of the mentors who helped guide him through his graduate and undergraduate studies of useful and harmful plants, the late Richard Evans Schultes at Harvard University and Richard W. Lighty at the University of Delaware. He also wishes to acknowledge the support of the Philecology Trust and the MetLife Foundation, through his appointment as a MetLife Fellow.

Finally, we thank our families for their patience and support during the research and writing of this book. Richard Shih wishes to specifically thank Laura, Catherine, Randy, Anne, Helen, and Chi Kai. Lewis Nelson is grateful to Laura, Daniel, Adina, and Benjamin as well as Myrna and Irwin. Michael Balick thanks Daniel and Tammy Balick and Roberta Lee. They have each given us the most precious contribution—time and understanding—which ultimately resulted in the volume you have before you. We hope you will find it worthy of their sacrifice.

Lewis S. Nelson, M.D.
Richard D. Shih, M.D.
Michael J. Balick, Ph.D.

Introduction

I studied botany before I studied medicine, having had the good fortune to pursue an undergraduate degree under the direction of the late Dr. Richard Evans Schultes, longtime director of the Harvard Botanical Museum and godfather of modern ethnobotany. Schultes was an expert on psychoactive and toxic plants, especially of the New World tropics. Initially, through his stories of the indigenous lifestyle of Amazonian peoples, and later by helping me undertake fieldwork in this region, he awoke in me a keen interest in the botany of useful plants that led me to become first an investigator and later a practitioner of botanical medicine.

When I moved on to Harvard Medical School, I was dismayed to find that none of my teachers, even of pharmacology, had firsthand knowledge of the plant sources of drugs. Since then I have been continually struck by the lack of awareness of the medicinal and toxic properties of plants in our culture. Examples are unfounded fears of poisoning by common ornamentals such as the poinsettia, exaggerated fears of herbal remedies such as Chinese ephedra, ignorance of the vast medicinal importance of such spices as turmeric and ginger, and lack of awareness of the toxic and psychoactive properties of other spices, for example, nutmeg and mace.

At the root of this problem is the distance that exists between plant scientists and health scientists. Because I am trained in both worlds, I have been very conscious of it all my professional life. This intellectual gap creates difficulties for botanists who want to learn the medical significance of plants with pharmacological effects and for physicians, nurses, and pharmacists who want to learn how plants influence health, whether for good or ill.

By bringing together specialists from both sides of this divide, the present book does a great service. It gives different perspectives on poisonous and injurious plants while remaining grounded in the integrative science of modern ethnobotany. I wish it had been available when I was first practicing medicine and, because of my background in botany, was often asked questions about the harmful potentials of plants and products derived from them.

I meet many people who imagine that most wild plants are dangerous, who think that if you pick and eat plants at random in the backyard or woods you will die. In fact, the percentage of plants that are really harmful is quite small, as is the percentage that are really beneficial. If you wish to get to know plants, a good place to start is to learn about those that can kill or cause serious harm. This handbook will be an invaluable resource in that educational process.

Andrew Weil, M.D.
Director, Program in Integrative Medicine
Clinical Professor of Medicine
University of Arizona Health Sciences Center
Tucson, Arizona, USA

Authors' Note

Poison Control Centers across the United States received more than 57,000 calls relating to plant exposure in 2003 (Watson et al., 2004), of which more than 85% involved children under the age of 6 years. Plant exposures account for the seventh most common form of reported toxic exposure in children (Table 1). This demographic is consistent with the ready availability of plants at home and in public locations and suggests that most plant exposures are unintentional. Similarly, the vast majority of these exposures result in no toxicity, an important fact that should be both settling and troublesome. Although it is likely that the majority of these plant exposures (Table 2) are nontoxic, it is certainly likely that most of these "exposures" were simply that—exposures. That is, no toxin was ingested, or if a small piece of plant was ingested, it was in a quantity insufficient to cause problems. However, the possibility of disregarding as nontoxic the rare patient with a substantial exposure is ever present. For this reason, a comprehensive understanding of the types of toxins present in a plant and the likely clinical manifestations following exposure is critical and the focus of this book.

Specific identification of a plant may guide management by revealing potential toxins, placing the risk in context, and providing a time frame for the development of clinical findings. Care should be taken to avoid misidentification, a particular problem when plants are discussed by their common rather than by their botanical name. Although management of a patient with an identified exposure is generally preferable to managing a patient with an "exposure to an unknown plant," many plant-exposed cases are managed successfully without knowledge of the culprit plant. However, adverse events may result by the attempted management of a misidentified plant. Each plant description in Section 5 is accompanied by one or more photographs to help the user of the book to qualitatively and tentatively identify an implicated plant (as well to provide a visual cue to those using the book as a learning tool). Not every specific plant discussed in the book is illustrated, nor are all the horticultural varieties illustrated. Some groups of ornamental plants may have hundreds of cultivars that have been named, each with a slightly different appearance or characteristic. Thus, many of the plants shown in the photographs are representative of the appearance of only a small group of species cultivars within the family or genus that might have toxic properties. The assistance of management

TABLE 1. Substances Most Frequently Involved in Pediatric Exposures (Children Under 6 Years) in 2003

Substance	Number	Percent
Cosmetics and personal care products	166,874	13.4
Cleaning substances	121,048	9.7
Analgesics	97,463	7.8
Foreign bodies	92,166	7.4
Topicals	92,091	7.4
Cough and cold preparations	68,493	5.5
Plants	**57,778**	**4.6**
Pesticides	50,938	4.1
Vitamins	45,352	3.6
Antimicrobials	35,152	2.8
Antihistamines	32,622	2.6
Arts/crafts/office supplies	31,211	2.5
Gastrointestinal preparations	29,770	2.4
Hormones and hormone antagonists	23,787	1.9
Electrolytes and minerals	22,337	1.8

Data from Watson et al. (2004)

TABLE 2. Frequency of Plant Exposures by Plant Type in 2003

Botanical name	Common name	Frequency
Spathiphyllum spp.	Peace lily	3,602
Philodendron spp.	Philodendron	2,880
Euphorbia pulcherrima	Poinsettia	2,620
Ilex spp.	Holly	2,427
Phytolacca americana	Pokeweed, inkberry	1,863
Ficus spp.	Rubber tree, weeping fig	1,612
Toxicodendron radicans	Poison ivy	1,500
Dieffenbachia spp.	Dumbcane	1,324
Crassula spp.	Jade plant	1,146
Epipremnum aureum	Pothos, devil's ivy	1,083
Capsicum annuum	Pepper	1,049
Rhododendron spp.	Rhododendron, azalea	1,047
Chrysanthemum spp.	Chrysanthemum	869
Nerium oleander	Oleander	847
Schlumbergera bridgesii	Christmas cactus	841
Hedera helix	English ivy	769
Eucalyptus spp.	Eucalyptus	727
Malus spp.	Apple, crabapple(plant parts)	703
Nandina domestica	Heavenly bamboo	694
Saintpaulia ionantha	African violet	685

Note: This table provides the frequency of involvement of plants in exposures reported to poison centers. These data do not imply actual exposure, poisoning, or any judgment with regard to toxicity. Several of the plants on the list pose little, if any, ingestion hazard.

Data from Watson et al. (2004)

algorithms and of books that help in plant identification is always appreciated, although this is unlikely to replace the assistance of a trained professional who is able to correctly identify plants. This person may typically be a professional botanist or a horticulturist, although some nurseries (Rondeau et al., 1992), which are more readily available, may have adequate expertise, particularly for common plants. A positive identification of an individual plant is most likely when a freshly collected part of the plant containing leaves and flowers or fruits is presented to the knowledgeable botanist or horticulturist. Poison Control Centers generally have relationships with the botanical community should the need for plant identification arise. Section 1 (Botanical Nomenclature and Glossary of Botanical Terms) provides an overview of botanical terms to ensure that the interaction between the botanical and medical communities is clear and efficient; this is critical to ensuring both safe and timely communication to meet the exacting demands of a clinical situation.

As most exposures result in little or no toxicity, the initial management of most incidents involving children who are asymptomatic should be expectant. This approach includes observation, at home or in the hospital as appropriate, depending on the nature of the exposure, and supportive care. For example, patients with several episodes of vomiting may benefit from an antiemetic agent and oral rehydration or, occasionally, intravenous fluids. Perhaps the greatest paradigm shift since the publication of the earlier edition of this book is the current deemphasis of aggressive gastrointestinal decontamination (see Section 4). Syrup of ipecac, for example, is almost never recommended, and orogastric lavage should be reserved for those patients with a reasonable likelihood of developing consequential poisoning. This group should include the minority of patients exposed to plants. Although oral activated charcoal is effective at reducing the absorption of many chemicals, its benefit following the vast majority of plant exposures has never been specifically studied. However, given the extremely low risk of administration of oral activated charcoal to an awake patient who is able to drink spontaneously, its use should be considered in patients with plant exposures. For complete information on the initial decontamination of the poisoned patient, call your regional Poison Control Center or refer to a textbook of medical toxicology, emergency medicine, or pediatrics.

Sections 2 ("Poisons, Poisoning Syndromes, and Their Clinical Management"), 3 (Plant-Induced Dermatitis [Phytodermatitis]), and 4 (Gastrointestinal Decontamination) include descriptions of the clinical findings and focused descriptions of management strategies for patients with plant poisonings. Although very few antidotes are available to treat the effects of the innumerable toxins available in plants, rarely are antidotes actually necessary. Much of our understanding of poisoning syndromes derives from toxicity associated with the use of purified plant toxins as pharmaceuticals (e.g., morphine from *Papaver somniferum*). The amount of a toxin present in a plant is unpredictable, whereas the amount in a tablet is always defined. There is generally a lower concentration of "toxin" in the plant than there is of "drug" in a tablet. However, this by no means should minimize the clinical concern following exposure to a plant containing a consequential toxin, such as *Colchicum autumnale*, which contains colchicine.

As already suggested, there is little adequate evidence to precisely direct the management of any specific plant poisoning. The limited knowledge relates to the wide diversity of available plants and the limited quality of available case data (e.g., did they eat it?). The cost and effort associated with proving an exposure (e.g., toxin levels in blood) makes this task (unfortunately but appropriately) of low priority to the physician involved with the care of the exposed patient. As with many other clinical situations, bedside care of patients with toxic plant exposures should be managed primarily based on their clinical manifestations and responses to therapy and only secondarily on the basis of the toxin to which they are presumably exposed. The dictum has been and remains "Treat the patient, not the poison". . . . but don't ignore the poison.

References

Rondeau ES, Everson GW, Savage W, Rondeau JH. Plant nurseries: A reliable resource for plant identification? Vet Hum Toxicol 1992;34:544–546.

Watson WA, Litovitz TL, Klein-Schwartz W, et al. 2003 Annual report of the American Association of Poison Control Centers Toxic Exposure Surveillance System. Am J Emerg Med 2004;22:335–404.

SECTION 1.

Botanical Nomenclature and Glossary of Botanical Terms

Botanical Nomenclature

Before the work of Carolus Linnaeus (1707–1778), the botanist who established the binomial system of plant nomenclature, a plant sometimes had a name that consisted of many descriptive words. Linnaeus helped to standardize botanical nomenclature by establishing a genus and species name for each plant, followed by its designator. A clinical report involving a plant must always include the plant's botanical (binomial) name, which consists of both the genus and the species, for example, *Duranta repens*. By convention, both are italicized or underlined. *Duranta* is the name of the genus and the first letter is always capitalized. A genus (the plural of which is genera) may be composed of a single species or several hundred. The second part of the binomial, in this case *repens*, is the particular species within the genus, and it is always in lowercase letters. It is important to include the name of the person (often abbreviated) who named the particular species, as part of the scientific name, to minimize confusion between similar or related plant species. For example, in the case above, the complete name, which would allow the most precise identification, is *Duranta repens* L.; L. is the accepted abbreviation for Carolus Linnaeus.

Over time, as botanists continue to revise the classification systems of their specific plant families or groups to reflect additional knowledge and a more natural, evolutionarily based system, plants are periodically moved into different genera or sometimes families. A species may be split into several species or varieties, or lumped together with plants of other species to comprise a single species, all based on the expertise of the taxonomist utilizing characteristics from other specialties ranging from gross morphology to molecular biology. One shortcoming of this fluid system is that scientists can have differing opinions as to how to classify a specific plant. To limit confusion with regard to nomenclature, when previously employed names are changed as part of a more recent taxonomic study, they become recognized as synonyms. In this book, the most common current synonyms are included in parentheses with an equal sign, for example, *Duranta repens* L. (= *D. plumferi* Jacq.). Some species are divided further into subspecies (ssp.), varieties (var.), cultivated varieties (cultivars (cv.)), and forms (fo.); for example, *Philodendron scandens* C. Koch & H. Sello ssp. *oxycardium* (Schott) Bunt. In this instance, the plant was first named *Philodendron oxycardium* by Heinrich Schott, but was reevaluated and then transferred to become a subspecies of *Philodendron scandens* by George Bunting. Hybrid names are indicated by an × (multiplication symbol), as in *Brugmansia × candida*. Horticultural names are not italicized but are capitalized and set in single quotation marks, for example, *Ilex glabra* cv. 'Compacta.' A printed work can never be fully up to date from a taxonomic standpoint because taxonomists are constantly refining the classification systems of the groups on which they work. At the same time, there may be a significant volume of medical literature based on an "older" name, and thus, for most efficient and rapid use of the information in this volume, some of the older names used in the first edition are retained.

Associations of like genera are placed in a family. The family name is not italicized, but the initial letter is always capitalized. Botanists have changed the status of some families to reflect a more natural evolutionary lineage, either by incorporating them into other families and dropping their original designation or by creating entirely new families. Since the publication of the original edition of this *Handbook*, family names for some of the genera have been changed, but in this new edition the older name has been maintained to facilitate rapid consultation of the toxicological literature, and the new name is added in parentheses, for example, Umbelliferae (= Apiaceae). We also head many of the poisoning syndromes in Section 2 with the name of the genus followed by the word "species" (spp.) to indicate that there are several to many species in this genus having toxic properties.

If an individual species cannot be found, but the genus is listed, it should be assumed, conservatively, that the species has a potential for toxicity similar to another member of that genus. To a lesser extent, such an association may exist for members of the same family (Table 3). These relationships are far from exact, and inconsistencies in the clinical presentation or therapeutic response of an exposed patient should prompt immediate consultation with a Poison Control Center or other expert source. The botanical nomenclature used in this book has been derived from various sources, as well as the opinions of specialist reviewers.

There are no rules for establishing common names of plants. Common names can be highly misleading and may erroneously suggest toxicity or the lack of toxicity. For example, a plant known as a "pepper" plant could be the sweet pepper commonly eaten as a vegetable (*Capsicum annuum* L. var. *annuum*); or one of the extremely hot, virtually "inedible" peppers (particularly when eaten in quantity and certainly depending on the person's palate) used as a decorative houseplant in that same species but containing significant quantities of capsaicin; or the spice plant from which we derive black pepper (*Piper nigrum*); or the pepper bush (*Leucothoe* species) containing grayanotoxins; or the pepper tree (*Schinus molle*) with triterpene-containing berries; or any number of other species with "pepper" as part of its common name. Another

TABLE 3. Examples of Plants Producing Systemic Poisoning in Humans Arranged by Family and Genus

Amaryllidaceae
Amaryllis
Hippeastrum
Clivia
Crinum
Galanthus
Hymenocallis
Lycoris
Narcissus
Zephyranthes

Anacardiaceae
Schinus

Apocynaceae
Acokanthera
Adenium
Allamanda
Nerium
Pentalinon
Thevetia

Aquifoliaceae
Ilex

Araceae
Arum
Alocasia
Anthurium
Arisaema
Caladium
Calla
Colocasia
Dieffenbachia
Epipremnum
Raphidophora
Monstera
Philodendron
Spathiphyllum
Symplocarpus
Xanthosoma
Zantedeschia

Araliaceae
Hedera

Asclepiadaceae
Calotropis
Cryptostegia

Berberidaceae
Caulophyllum
Podophyllum

Boraginaceae
Echium
Heliotropium

Calycanthaceae
Calycanthus

Campanulaceae
Hippobroma
Lobelia

Caprifoliaceae
Lonicera
Sambucus
Symphoricarpos

Celastraceae
Celastrus
Euonymus

Compositae
Senecio

Coriariaceae
Coriaria

Cornaceae
Aucuba

Corynocarpaceae
Corynocarpus

Cucurbitaceae
Momordica

Cycadaceae
Cycas

Ericaceae
Kalmia
Leucothoe
Lyonia
Pernettya
Pieris
Rhododendron

Euphorbiaceae
Aleurites
Euphorbia
Hippomane
Hura
Jatropha
Manihot
Pedilanthus
Ricinus

Ginkgoaceae
Ginkgo

Guttiferae
Calophyllum
Clusia

Hippocastanaceae
Aesculus

Iridaceae
Iris

Leguminosae
Abrus
Baptisia
Caesalpinia
Cassia
Crotalaria
Gymnocladus
Laburnum
Leucaena
Pachyrhizus
Robinia
Sesbania
Sophora
Wisteria

Lilliaceae
Allium
Aloe
Bulbocodium
Colchicum
Convallaria
Gloriosa
Ornithogalum
Schoenocaulon
Scilla
Urginea
Veratrum
Zigadenus

TABLE 3. *Continued*

Loganiaceae	**Ranunculaceae**	**Solanaceae**
Gelsemium	*Aconitum*	*Atropa*
Spigelia	*Actaea*	*Capsicum*
Strychnos	*Adonis*	*Cestrum*
	Anemone	*Datura*
Loranthaceae	*Caltha*	*Brugmansia*
Phoradendron	*Clematis*	*Hyoscyamus*
Viscum	*Helleborus*	*Lycium*
	Pulsatilla	*Nicotiana*
Meliaceae	*Ranunculus*	*Physalis*
Melia		*Solandra*
Swietenia	**Rhamnaceae**	*Solanum*
	Karwinskia	
Menispermaceae	*Rhamnus*	**Taxaceae**
Menispermum		*Taxus*
	Rosaceae	
Myoporaceae	*Eriobotrya*	**Thymelaeaceae**
Myoporum	*Malus*	*Daphne*
	Prunus	*Dirca*
Oleaceae	*Rhodotypos*	
Ligustrum		**Umbelliferae**
	Rutaceae	*Aethusa*
Palmae	*Poncirus*	*Cicuta*
Caryota		*Conium*
	Sapindaceae	*Oenanthe*
Papaveraceae	*Blighia*	
Chelidonium	*Sapindus*	**Verbenaceae**
		Duranta
Phytolaccaceae	**Saxifragaceae**	*Lantana*
Phytolacca	*Hydrangea*	
Rivina		**Zamiaceae**
	Scrophulariaceae	*Zamia*
Polygonaceae	*Digitalis*	
Rheum		

problem associated with common names is that they can sometimes lead to the assumption that plants are related—either botanically or toxicologically. For example the "hellebore," *Helleborus niger* L., is in the family Ranunculaceae, but it bears no relationship to the "false hellebore," *Veratrum viride* Aiton, a member of the family Liliaceae; the former species contains toxic glycosides and saponins and the latter contains toxic alkaloids. The botanical (binomial) nomenclature is essential for ensuring proper plant identification.

Common names are included throughout this book only to facilitate in the identification of a particular plant in question. Many common names are no longer in use and others have been developed, but there is no way to verify contemporary use except by interviewing the inhabitants of a region and recording their responses. Thus, for a compilation of common names in this text we depended on the literature. The common names of native species from the United States and Canada are taken from Kartesz and Kartesz (1980). Names for West Indian species and Guam were selected from the floras listed in the references. Common names for cultivated plants were taken primarily from *Hortus Third*. In addition to floras, Hawaiian names are from Neal (1965), Cuban names from Roig y Mesa (1953), and Mexican names from Aguilar and Zolla (1982). Many less-common, older names for plants in the United States were selected from Clute (1940). When bolded, the common name connotes the most widely employed name in contemporary use in the United States.

Care must be exercised when evaluating poisonous plant literature. In some instances, information on the toxicity of plants in grazing animals is extrapolated to predict that which may occur in humans. Unsubstantiated plant lore has passed through generations of textbooks; we have attempted to remove as much lore as possible. Even evaluations based on human case reports, which act as the foundation for this book, may be flawed by erroneous identification of the plants or inappropriate attribution of the clinical effects to the plant.

Glossary of Botanical Terms

This list of botanical and horticultural terms is provided to aid in understanding the plant descriptions found in the text. The terms have mostly been taken from two primary references, *Manual of Vascular Plants of Northeastern United States and Adjacent Canada,* Second Edition (Henry A. Gleason and Arthur Cronquist, 1991) and *Hortus Third: A Concise Dictionary of Plants Cultivated in the United States and Canada* (Liberty Hyde Bailey and Ethel Zoe Bailey, Revised and Expanded by The Staff of the Liberty Hyde Bailey Hortorium, 1976). Some definitions have been modified from the original for ease of use and understanding by the nonbotanist, and the reader is urged to consult a botanical textbook if greater detail is required. The botanical illustrations are by Bobbi Angell.

Alternate: Arranged singly at different heights and on different sides of the stem— as in alternate leaves.

Annual: Yearly; a plant that germinates, flowers, and sets seed during a single growing season.

Anther: The portion of the stamen of a flower that contains the pollen, usually having two connected pollen sacs.

Aril: A specialized, usually fleshy outgrowth that is attached to the mature seed; more loosely, any appendage or thickening of the seed coat.

Bark: Outer surface of the trunk of a tree or woody shrub.

Bearded: Bearing a tuft or ring of rather long hairs.

Berry: The most generalized type of fleshy fruit, derived from a single pistil, fleshy throughout, and containing usually several or many seeds; more loosely, any pulpy or juicy fruit.

Biennial: Living 2 years only and blooming the second year.

Blade: The expanded, terminal portion of a flat organ such as a leaf, petal, or sepal, in contrast to the narrowed basal portion.

Bony: Hard surface as in a bone.

Bract: Any more or less reduced or modified leaf associated with a flower or an inflorescence that is not part of the flower itself.

Bulbil, bulblet: Diminutive of bulb; one of the small new bulbs arising around the parent bulb; a bulblike structure produced by some plants in the axils of leaves or in place of flowers.

Bulb: A short vertical, underground shoot that has modified leaves or thickened leaf bases prominently developed as food-storage organs.

bulblet   bulb

Buttress: Flattened support structures at the base of the trunk of certain types of trees, particularly in the tropics.

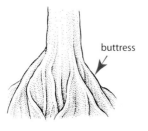

buttress

Calyx: All the sepals of a flower, collectively.

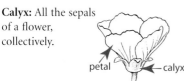

petal   calyx

Capsule: A dry, dehiscent fruit composed of more than one carpel.

Carpel: The fertile leaf of an angiosperm that bears the ovules. The pistil (female part of the flower) is made up of one or more carpels, where the seeds normally are found.

Climbing: Growing more or less erect without fully supporting its own weight, instead leaning, scrambling, twining, or attaching onto some other structure such as a tree or wall.

Coarse: Rough, as in the texture of a leaf.

Compound leaf: A leaf with two or more distinct leaflets.

leaflet

Cone: A cluster of sporophylls or ovuliferous scales on an axis; a strobilus, as in pine or cycad cones.

Corolla: All the petals of a flower collectively.

Corona: A set of petal-like structures or appendages between the corolla and the androecium (male element of the flower).

corona

Creeping: Growing along (or beneath) the surface of the ground and rooting at intervals, usually at the nodes.

Cultivar: A horticultural variety originating from a cultivated plant, possessing interesting or important characters such as color, smell, taste, or disease resistance that make it worthy of distinction through naming.

Cuttings: Small pieces of stems or roots that can be put in soil to develop into a complete plant.

Cyme: A broad class of inflorescences characterized by having the terminal flower bloom first, commonly also with the terminal flower of each branch blooming before the others on that branch.

Deciduous: Falling after completion of the normal function. A deciduous tree is one that normally loses its leaves at the approach of winter or the dormant season.

Dehiscent: Opening when mature, exposing or releasing the contents, as in a fruit releasing it seeds.

Dicotyledons: One of the two major divisions of the angiosperms (a group characterized by having ovules borne in ovaries) bearing two (or rarely more) cotyledons or seed leaves, comprising most of the familiar seed plants.

Divided: Cut into distinct parts, as a leaf that is cut to the midrib or the base.

Drupe: A fleshy fruit with a firm endocarp ("pit" or "stone") that permanently encloses the usually solitary seed, or with a portion of the endocarp separately enclosing each of two or more seeds.

Ellipsoid: Elliptical in long section and circular in cross section (applied only to three-dimensional bodies).

Elliptic: With approximately the shape of a geometric ellipse (applied only to flat bodies).

Erect: Upright.

Escaped: As in an introduced plant species that has escaped from cultivation into the wild.

Evergreen: Remaining green throughout the winter, as in a tree that keeps its leaves throughout the year.

Feathery: Feather shaped in outline, as in leaves.

Female flowers: Referring to flowers that are pistillate, having pistils but no stamens.

Filament: The stalk of a stamen, that is, the part that supports the anther.

Finely toothed leaves: Leaves with small serrations on the edges.

Fishtail-shaped: As in leaflets of some palms that have a somewhat irregularly triangular or "fishtail" outline.

Fleshy: Thick and juicy; succulent.

Floral bracts: Greatly reduced leaf associated with a flower, usually at its base.

Floral branches: Branches or axes on which flowers are formed.

Flower: An axis bearing one or more pistils or one or more stamens or both.

Fruit: A ripened ovary along with any other structures that may ripen with it and form a unit with it.

Fruit pulp: Fleshy material inside of a fruit, often the part that is eaten by humans or animals.

Funnel-form: Shaped like a funnel, as in a flower.

Furrowed (stems): Having longitudinal channels or grooves along the stem.

Glaucous: Covered with a fine, waxy, removable powder that imparts a whitish or bluish cast to the surface, as in a prune or a cabbage leaf.

Globose: More or less spherical.

Glossy: Shiny.

Head: A cluster of flowers crowded closely together at the tip of a floral stem.

Herb: A plant, either annual, biennial, or perennial, with the stems dying back to the ground at the end of the growing season, and without woody stems.

Herbaceous: Adjectival form of herb; also, leaflike in color or texture, or not woody.

Hilum: The scar of the seed at its point of attachment.

hilum

Horticultural varieties: As in cultivars.

Hybrid: A plant that results from a cross between two parent species that are genetically different.

Indehiscent: Remaining closed at maturity.

Inflorescence: A flower cluster of a plant; the arrangement of the flowers on the axis.

Juvenile leaves: A younger form or shape of the leaves of a plant, which change when the plant reaches maturity.

Lacy leaves: As in the shape of leaves with many tears or cuts.

Lance-shaped: As in leaves that are several times longer than broad and widest below the middle, tapering with convex sides upward to the tip.

Latex: A colorless, white, yellow, or reddish liquid, produced by some plants, characterized by the presence of colloidal particles of terpenes dispersed in water.

Leaflet: An ultimate unit of a compound leaf. (see **Compound leaf**)

Leathery: Thick and leatherlike in texture, as in a leaf.

Lobe: A projecting segment of an organ, too large to be called a tooth, but with the adjoining sinuses usually extending less than halfway to the base or midline.

Mature fruit: A fruit that has ripened; often a different color from when it was young.

Midrib: The main rib or longitudinal vein (an externally visible vascular bundle) of a leaf or leaflet.

Milky latex: White-colored sap of a plant.

Monocotyledons: One of two major divisions of the angiosperms (a group of plants characterized by having ovules borne in ovaries), bearing only one cotyledon or seed leaf, for example, the grasses, lilies, bromeliads, orchids, and palms.

Native: Having its origins in a particular geographic area, as in a plant native to the Western United States.

Naturalized: Thoroughly established in a particular geographic region, but originally coming from another geographic area.

New World: Pertaining to the Western Hemisphere, particularly the Americas, as in a plant native to that region.

Nut: A relatively large, dry, indehiscent fruit with a hard wall, usually containing only one seed.

Oblong: Shaped more or less like a geometric rectangle (other than a square).

Obovate: Similar to ovate but larger toward the tip of the leaf.

Old World: Pertaining to Europe, Asia, and Africa, as in a plant native to that region.

Opposite: Situated directly across from each other at the same node or level, as in the leaves or leaflets of some plants; situated directly in front of (on the same radius as) another organ, as stamens opposite the petals.

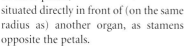

Ovate: Shaped like a long section through a hen's egg, with the larger end toward the base.

Ovule: A young or undeveloped seed.

Palmately compound: As in a leaf with three or more lobes arising from a common point.

Panicle: A branching indeterminate inflorescence, usually broadest near the base and tapering upwards.

Pantropical: Found throughout the tropical regions.

Pedicellate: Borne on a pedicel (stalk of a single flower in an inflorescence).

Pendant: Hanging, as in pendant racemes of flowers.

Perennial: A plant living more than 2 years.

Petal: A member of the inner set of floral leaves, usually colored or white and serving to attract pollinators.

Pistil: The female organ of a flower, ordinarily differentiated into an ovary, style, and stigma.

Pit: Hardened covering enclosing seed or seeds in a fruit, as in a peach.

Pleated: When young, as in a leaf, folded several times along the length.

Pod: Any kind of dry, dehiscent fruit.

Prickle: A sharp outgrowth from the epidermis or bark.

Propagated: As in multiplying a plant through making cuttings and planting them.

Pubescent: Bearing hairs (trichomes) of any sort.

Raceme: A more or less elongate inflorescence with pedicellate flowers arising in a sequence from the bottom up from an unbranched central axis.

Recurved petals: Flower petals that are curved downward or backward.

Resinous: Containing resin.

Rhizome: A creeping underground stem.

Rosette: A cluster of leaves or other organs arranged in a circle or disk, often in a basal position.

Runner: A long, slender, prostrate stem rooting at the nodes and tip.

Sap: Liquid contained within the stem.

Scale: Any small, thin, or flat structure.

Scaly: Covered with scales or bracts.

Scorpioid cyme: A coiled inflorescence with flowers developing alternately to left and right in a zigzag fashion.

Seed coat: Outside coating of a seed.

Seedpods: As in a fruit or pod containing seeds.

Sepal: One of the outermost set of floval leaves. (see **Calyx**)

Serrate: Toothed along the margin with sharp, forward-pointing teeth.

Serrated leaf: Saw toothed, with teeth pointing forward toward the tip of the leaf.

Showy: Conspicuous and ornamental.

Shrub: A woody plant that remains low and produces shoots or trunks from its base.

Silky: A covering of fine, soft hairs.

Simple leaf: A leaf with the blade all in one piece (although it may be deeply cleft), not compound.

Sinus: The cleft between two lobes or segments of a leaf.

Spadix: The thick or fleshy flower spike of certain plants, usually surrounded by or subtended by a bract, as in the Araceae.

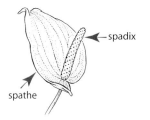

Spathe: A large, usually solitary bract subtending and often enclosing an inflorescence; the term is used only in the monocotyledons.

Spearhead-shaped: As in a leaf shaped like the head of a spear.

Spike: A more or less elongate inflorescence, with sessile (lacking a stalk) flowers attached directly by their base.

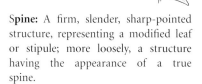

Spine: A firm, slender, sharp-pointed structure, representing a modified leaf or stipule; more loosely, a structure having the appearance of a true spine.

Sporophyll: A modified leaf that bears or subtends the spore-bearing cases in certain plants such as ferns and cycads.

Sprays: Clusters of flowers.

Stamen: The male organ of a flower, consisting of an anther usually on a filament.

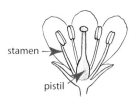

Stipule: One of a pair of basal appendages found on the leaf petiole of many species.

Strap-shaped, straplike: As in a long, narrow, thick leaf.

Strobilus: A cluster of sporophylls or ovule-bearing scales on an axis, such as in a cone.

Tendril: A slender, coiling, or twining organ (representing a modified stem or leaf or part thereof) by which a climbing plant grasps its support.

Terminal clusters: As in flowers clustered at the end or tip of a branch.

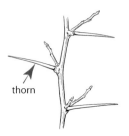

thorn

Thorn: A stiff, woody, modified stem with a sharp point.

Throat: The opening or orifice of a fused corolla or calyx, or the somewhat expanded part between the proper tube and the limb; in grasses, the upper margins of the sheath.

Tooth: Serration, as on the edge of a leaf (plural, teeth).

Tuberous: Thickened like a tuber, as in roots.

Variegated: Differently colored areas, as in a leaf with streaks, marks, or patterns of various colors on its surface.

Variety: A subdivision of a species ranking lower than a subspecies.

Velvety: With erect, straight, moderately firm hairs, such as on a stem or leaf.

Warty: Covered with wartlike structures.

Weed: A plant that aggressively colonizes disturbed habitats or places where it is not wanted.

Winged fruit/seed: A thin, flat extension or projection from the side or tip of a seed.

Botanical References for This Volume

Adams CD. *Flowering Plants of Jamaica*. University of West Indies, Mona, Jamaica, 1972.

Aguilar Contreras A, Zolla C. *Plantas Tóxicas de México*. Subdirección General Médica, División de Información Ethnobotánica, Unidad de Investigación Biomédica en Medicina Tradicional y Herbolaria del Instituto Mexicano del Seguro Social, Mexico D.F., México, 1982.

Bailey LH, Bailey EZ, Liberty Hyde Bailey Hortorium Staff. *Hortus Third*. Macmillan, New York, 1976.

Barker HD, Dardeau WS. *Flore d'Haïti*. Publié sous la direction du Service technique du Département de l'agriculture et de l'enseignement professionnel. Port-au-Prince, Haiti, 1930.

Clute WN. *American Plant Names*, 3rd ed. Willard N. Clute, Indianapolis, 1940.

Correll DS, Correll HB. *Flora of the Bahama Archipelago*. Vaduz J. Cramer, Germany, 1982.

Gleason HA, Cronquist A. *Manual of Vascular Plants of Northeastern United States and Adjacent Canada*, 2nd ed. The New York Botanical Garden, New York, 2004.

Gooding EGB, Loveless AR, Proctor GR. *Flora of Barbados*. Her Majesty's Stationery Office, London, 1965.

Howard RA (ed) *Flora of the Lesser Antilles: Leeward and Windward Islands*. Arnold Arboretum, Harvard University, Jamaica Plain, Massachusetts, 1974.

Kartesz JT, Kartesz R. *A Synonymized Checklist of the Vascular Flora of the United States, Canada, and Greenland*. University of North Carolina Press, Chapel Hill, 1980.

Kuijt J. *Monograph of Phoradendron* (*Viscaceae*). Systematic Botany Monographs, vol 66. The American Society of Plant Taxonomists, University of Wyoming, Laramie, 2003.

León Hermano (Sauget J). 1946. *Flora de Cuba*, vol 1. Museum de Historia Natural del Collegio de La Salle, Havana, 1957.

León Hermano (Sauget J), Hermano Alain (Liogier H). *Flora de Cuba*, vols 2–4. Museum de Historia Natural del Collegio de La Salle, Havana, 1951–1957.

Liogier AH. *Flora de Cuba*, vol 5. Editorial Universitario, Universidad de Puerto Rico, Rio Piedras, Puerto Rico, 1962.

Liogier AH. *Flora de Cuba, Supplemento*. Editorial Sucre, Caracas, 1969.

Liogier AH. *Diccionario Botánico de Nombres Vulgares de la Español*. Jardín Botáníco Dr. Rafael Moscoso, Santo Domingo, 1974.

Liogier AH, Martorell LF. *Flora of Puerto Rico and Adjacent Islands: A Systematic Synopsis*. Rio Piedras, P.R., Editorial de la Universidad de Puerto Rico, 1982.

Neal MC. *In Gardens of Hawaii*. Bishop Museum Press, Honolulu, 1965.

Roig y Mesa JT. *Diccionario Botanico de Nombres Vulgares Cubanos.* Bulletin 45. Ministry of Agriculture, Havana, 1953.

Scoggan HJ. *Flora of Canada* (4 vols). National Museum of Natural Sciences, Ottawa, 1978–1979.

Stone BC. The flora of Guam. Micronesica 1970;6:1–659.

Tutin TG, Heywood VH, Burgess NA, Valentine DH, Walters SM, Webb DA. *Flora Europaea* (5 vols). Cambridge University Press, New York, 1964–1980.

Note: **Relevant medical references are included in the description of each plant. Several books were used as general references, particularly for less-common plant exposures. Although every attempt was made to provide literature-based support for our clinical and therapeutic descriptions, in certain instances we needed to rely on these older publications.**

Frohne D. *A Colour Atlas of Poisonous Plants: A Handbook for Pharmacists, Doctors, Toxicologists, and Biologists.* Wolfe, London, 1984.

Kingsbury JM. *Poisonous Plants of the United States and Canada*, 3rd ed. Prentice Hall, Englewood Cliffs, NJ, 1964.

Kingsbury JM. *Deadly Harvest: A Guide to Common Poisonous Plants.* Holt, Rinehart & Winston, New York, 1965.

Leopold WC. *Poisonous Plants of the United States.* Macmillan, New York, 1951.

SECTION 2.

Poisons, Poisoning Syndromes, and Their Clinical Management

This section includes detailed scientific and clinical toxicological information. For nonphysicians, this scientific and medical information should be utilized for informational purposes only; certain medical terms utilized in this section terms are not defined. For physicians, clinical care should not simply focus on the potential plant exposure, although this information is obviously helpful in many situations. Rather, clinical care should take into account the patient's current and prior medical history, physical examination, appropriate diagnostic testing, response to therapy, and all other factors normally utilized in the provision of clinical care. That is, the patient should be managed based on his or her clinical condition rather than just on the knowledge of an exposure or suspicion of a toxin. Appropriate clinical judgment should be exercised in the management of all patients. The information in this section should be supplemented by consultation with a Poison Control Center, medical toxicologist or other expert, or the use of a medical textbook or other appropriate reference.

General initial medical management strategies that are required for all plant-exposed patients include, but are not necessarily limited to, vital sign assessment, consideration of the need for immediate interventions (e.g., ventilation and oxygenation, blood glucose), determination of the need for laboratory or other diagnostic testing, and the consideration of the need for gastrointestinal decontamination (see Section 4). Intervention at any point that is deemed appropriate to correct or prevent progression of a clinical abnormality is critical. Specific considerations and interventions follow. Additional information and references are found in the individual plant descriptions in Section 5.

Poisoning by Plants with Anticholinergic (Antimuscarinic) Poisons
Examples of plant genera associated with this syndrome:

Atropa	*Brugmansia*	*Datura*
Hyoscyamus	*Solandra*	*Solanum*

Toxic Mechanism
Competitive antagonism of acetylcholine at the muscarinic subtype of the acetylcholine receptor, which is primarily located in the parasympathetic nervous system and the brain.

21

Clinical Manifestations

The classically described anticholinergic syndrome includes dry, warm, and flushed skin, parched mucous membranes, garbled speech, sinus tachycardia, adynamic ileus (absent bowel motility), urinary retention, and delirium with hallucinations. The hallucinations may be quite troubling to the patient, and patients may develop severe dysphoria or agitated delirium along with their sequelae. The patient's temperature may be slightly elevated, and is rarely above 102°F unless he or she is severely agitated or convulsing. Seizures may occur but generally only in patients who have other clinical findings consistent with anticholinergic poisoning. Complete clinical recovery even in the absence of complications may take many hours to days.

Specific Therapeutics

Given the common clinical presentation of altered mental status in association with elevated body temperature, patients should be evaluated for other medical problems, including sepsis and meningitis, unless the diagnosis is certain. Patients who are seriously poisoned by an antimuscarinic agent, particularly those with an appropriate confirmatory history, should receive either sedation with a benzodiazepine or reversal of their clinical syndrome with physostigmine. This antidote, a cholinesterase inhibitor, raises intrasynaptic levels of acetylcholine by preventing the neurotransmitter's enzymatic metabolism by the enzyme cholinesterase and allows acetylcholine to successfully compete with the toxin for the muscarinic receptor. The initial dose of physostigmine is 1–2 mg in adults (0.02 mg/kg in children) administered intravenously over no less than 5 minutes. Lack of clinical improvement suggests that either the diagnosis is incorrect or the dose of physostigmine is insufficient. Failure to develop cholinergic findings (e.g., salivation, bradycardia) following physostigmine raises the likelihood of the diagnosis of anticholinergic poisoning, and administration of increasing doses of the drug (up to 5 mg total dose in adults over 30 minutes) may be appropriate. The duration of action of some of the antimuscarinic alkaloids may be longer than that of physostigmine, and repeated administration of the latter may be required; alternatively, once the diagnosis is confirmed by an appropriate response to antidote, the patient may be sedated with a benzodiazepine and observed.

References

Burns MJ, Linden CH, Graudins A, Brown RM, Fletcher KE. A comparison of physostigmine and benzodiazepines for the treatment of anticholinergic poisoning. Ann Emerg Med 2000;35:374–81.
Howland MA. Physostigmine salicylate. In: Flomenbaum NE, Goldfrank LR, Hoffman RS, Howland MA, Lewin NA, Nelson LS. *Goldfrank's Toxicologic Emergencies, 8th Edition.* McGraw-Hill. New York, NY. 2006. p. 794–797.

Poisoning by Plants with Calcium Oxalate Crystals

Examples of plant genera associated with this syndrome:

Alocasia	*Arisaema*	*Brassaia*
Caladium	*Caryota*	*Colocasia*
Dieffenbachia	*Epipremnum*	*Monstera*
Philodendron	*Spathiphyllum*	

Calcium oxalate crystals at high magnification*

Toxic Mechanism

Upon mechanical stimulation, as occurs with chewing, crystalline calcium oxalate needles, bundled in needle-like raphides, release from their intracellular packaging (idioblasts) in a projectile fashion. These needles penetrate the mucous membranes and induce the release of histamine and other inflammatory mediators.

Clinical Manifestations

After biting or chewing, there is rapid onset of local oropharyngeal pain, which typically limits continued exposure, as well as local swelling and garbled speech. If swallowed, inflammation of the posterior oropharynx or larynx may rarely produce oropharyngeal edema and airway compromise. Endoscopic evaluation of the patient's airway, esophagus, or stomach may be necessary. Ocular exposure produces extreme pain, keratoconjunctival injection, and chemosis, with the potential for severe ocular damage and vision loss. Extensive dermal contact may produce pain and signs of irritation. In contrast to that occurring with soluble oxalate ingestion, in which profound hypocalcemia may occur, there is generally no associated systemic toxicity.

Specific Therapeutics

Airway assessment and management is of the highest priority following ingestion. Oropharyngeal or dermal pain may be managed with appropriate demulcents, viscous lidocaine, analgesics or with copious irrigation. Further evaluation of the patient's pharyngeal, respiratory, and gastrointestinal tract may be necessary. Eye exposure generally requires extensive irrigation and analgesia. Ophthalmologic consultation should be considered as needed.

References

* Franceschi VR, Nakata PA. Calcium oxalate in plants: Formation and function. Annu Rev Plant Biol. 2005;56:41–71.

Gardner DG. Injury to the oral mucous membranes caused by the common houseplant, Dieffenbachia. A review. Oral Surg Oral Med Oral Pathol. 1994;78:631–633.

Palmer M, Betz JM. Plants. In: Flomenbaum NE, Goldfrank LR, Hoffman RS, Howland MA, Lewin NA, Nelson LS. *Goldfrank's Toxicologic Emergencies, 8th Edition.* McGraw-Hill. New York, NY. 2006. p. 1577–1602.

Poisoning by Plants with Cardioactive Steroids/Cardiac Glycosides
Examples of plant genera associated with this syndrome:

Acokanthera	*Adenium*	*Adonis*
Calotropis	*Cryptostegia*	*Digitalis*
Helleborus	*Ornithogalum*	*Convallaria*
Nerium	*Pentalinon*	*Thevetia*
Urginea	*Strophanthus*	*Scilla*

Toxic Mechanism
Cardioactive steroids, termed cardiac glycosides when sugar moieties are attached, inhibit the cellular Na^+/K^+-ATPase. The effect is to indirectly increase intracellular Ca^{2+} concentrations in certain cells, particularly myocardial cells. Therapeutically, this both enhances cardiac inotropy (contractility) and slows the heart rate. However, excessive elevation of the intracellular Ca^{2+} also increases myocardial excitability, predisposing to the development of ventricular dysrhythmias. Enhanced vagal tone, mediated by the neurotransmitter acetylcholine, is common with poisoning by these agents, and produces bradycardia and heart block.

Clinical Manifestations
Ingestion of plants containing cardioactive steroids may cause abdominal pain and induce vomiting, which serves both as an early sign of toxicity and a mechanism to limit poisoning. Cardiovascular and electrocardiographic effects include sinus and junctional bradycardia as well as ventricular tachydysrhythmias, including ventricular tachycardia and ventricular fibrillation. Hyperkalemia may develop and is associated with poor patient outcome. Serum digoxin concentrations may be obtained but should not be relied upon to exclude toxicity as other cardioactive steroids will have unpredictable assay

cross-reactivity. Consequently, treatment, if clinically indicated, should not await laboratory confirmation.

Specific Therapeutics

Most of the available clinical experience with cardioactive steroid poisoning is related to digoxin toxicity. In these patients, standard supportive medical management is often inadequate. Therefore, any patient with consequential digoxin poisoning should receive digoxin-specific Fab. This product contains the antigen-binding regions (Fab) of animal-derived antidigoxin antibodies. Although specifically designed for the management of digoxin poisoning, digoxin-specific Fab appears to have sufficient cross-recognition of other cardioactive steroids to warrant its administration in other nondigoxin cardioactive steroid poisonings. The empiric dose is 10 vials (400 mg) administered intravenously in both adults and children, with additional dosing based on clinical response or additional information. Indications for its use include significant bradycardia, tachydysrhythmias, or hyperkalemia, with or without an elevated serum digoxin concentration, in any patient seriously believed to be poisoned by a cardioactive steroid-containing plant.

References

Eddleston M, Persson H. Acute plant poisoning and antitoxin antibodies. J Toxicol Clin Toxicol 2003;41:309–315.

Hack JB, Lewin NA. Cardioactive steroids. In: Flomenbaum NE, Goldfrank LR, Hoffman RS, Howland MA, Lewin NA, Nelson LS. *Goldfrank's Toxicologic Emergencies, 8th Edition.* McGraw-Hill. New York, NY. 2006. p. 971–982.

Howland MA. Digoxin-specific antibody fragments. In: Flomenbaum NE, Goldfrank LR, Hoffman RS, Howland MA, Lewin NA, Nelson LS. *Goldfrank's Toxicologic Emergencies, 8th Edition.* McGraw-Hill. New York, NY. 2006. p. 983–988.

Newman LS, Feinberg MW, LeWine HE. Clinical problem-solving: A bitter tale. N Engl J Med 2004;351:594–599.

Poisoning by Plants with Convulsant Poisons (Seizure)
Examples of plant genera associated with this syndrome:

Aethusa	*Anemone*	*Blighia*
Caltha	*Caulophyllum*	*Cicuta*
Clematis	*Conium*	*Coriaria*
Gymnocladus	*Hippobroma*	*Laburnum*
Lobelia	*Menispermum*	*Myoporum*
Nicotiana	*Pulsatilla*	*Ranunculus*
Sophora	*Spigelia*	*Strychnos*

Toxic Mechanism

A convulsion is the rhythmic, forceful contraction of the muscles, one cause of which are seizures. Seizures are disorganized discharges of the central nervous system that generally, but not always, result in a convulsion. There are various toxicological mechanisms that result in seizures including antagonism of gamma-aminobutyric acid (GABA) at its receptor on the neuronal chloride channel, imbalance of acetylcholine homeostasis, excitatory amino acid mimicry, sodium channel alteration, or hypoglycemia. Strychnine and its analogues antagonize the postsynaptic inhibiting activity of glycine at the spinal cord motor neuron. Strychnine results in hyperexcitability of the motor neurons, which manifests as a convulsion.

Clinical Manifestations

Unless an underlying central nervous system lesion exists, patients with plant-induced seizures generally present with generalized, as opposed to focal, seizures. Most patients develop generalized tonic-clonic convulsions, in which periods of shaking movement (convulsion or clonus) are interspersed with periods of hypertonicity. Occasionally, patients may not have overt motor activity (i.e., nonconvulsive seizure), or may present in the postictal period (partially or fully recovered from their seizure). The diagnosis in this situation may be difficult to determine. Patients who are having a generalized seizure should have loss of consciousness as a result of central nervous system dysfunction, and often have urinary or fecal incontinence, tongue biting, or other signs of trauma.

Conscious patients who are manifesting what appear to be generalized convulsions may have myoclonus or strychnine poisoning. Strychnine-poisoned patients manifest symmetrical convulsive activity, but because the activity is the result of spinal cord dysfunction, there is no loss of consciousness (i.e., no seizure) until metabolic or other complications intercede.

Specific Therapeutics

Once hypoglycemia and hypoxia have been excluded (or treated), a rapidly acting anticonvulsant benzodiazepine (e.g., diazepam, 5–10 mg in adults (0.1–0.3 mg/kg in children) or lorazepam, 2 mg in adults, or 0.1 mg/kg in children), should be administered parenterally for persistent seizures. Although diazepam and lorazepam are nearly equivalent in time to onset, lorazepam has a substantially longer duration of anticonvulsant effect. Lorazepam can be administered intramuscularly, though this route is not ideal because of a slow absorptive phase. Dosing may be repeated several times as needed. Inability to expeditiously control seizures with benzodiazepines may necessitate the administration of barbiturates, propofol, or another anticonvulsant medication. There is generally no acute role for phenytoin or other maintenance anticonvulsants in patients with toxin-induced seizures.

References

Chan YC. Strychnine. In: Flomenbaum NE, Goldfrank LR, Hoffman RS, Howland MA, Lewin NA, Nelson LS. *Goldfrank's Toxicologic Emergencies, 8th Edition.* McGraw-Hill. New York, NY. 2006. p. 1492–1496.

Philippe G, Angenot L, Tits M, Frederich M. About the toxicity of some *Strychnos* species and their alkaloids. Toxicon 2004;44:405–416.

Wills B, Erickson T. Drug- and toxin-associated seizures. Med Clin North Am 2005;89: 1297–1321.

Poisoning by Plants with Cyanogenic Compounds

Examples of plant genera associated with this syndrome:

Eriobotrya	*Hydrangea*	*Malus*
Prunus	*Sambucus*	

Toxic Mechanism

Cyanogenic compounds, most commonly glycosides, must be metabolized to release cyanide. Cyanide inhibits the final step of the mitochondrial electron transport chain, resulting rapidly in cellular energy failure.

Clinical Manifestations

Because the cyanogenic glycosides must be hydrolyzed in the gastrointestinal tract before cyanide ion is released, the onset of toxicity is commonly delayed. Abdominal pain, vomiting, lethargy, and sweating develop initially, followed shortly by altered mental status, seizures, cardiovascular collapse, and multisystem organ failure. Laboratory testing may reveal an elevated blood lactic acid; cyanide levels are not generally available rapidly. Thiocyanate, a metabolite of cyanide, may be measured in the patient's blood, and although often confirmatory in retrospect, immediate results are not readily available.

Specific Therapeutics

Initial management includes aggressive supportive care, intravenous fluid therapy, and correction of consequential metabolic acidosis using intravenous sodium bicarbonate as appropriate. Antidotal therapy, available in the form of a prepackaged cyanide antidote kit, should be administered to any patient believed to be suffering from cyanide poisoning. Before the establishment of an intravenous line, an amyl nitrite pearl may be broken and held under the patient's nose for 30 seconds each minute. In patients with intravenous access,

10 ml of 3% sodium nitrite in an adult, or in an appropriate pediatric dose (guidelines supplied with the kit), should be administered intravenously; this should be followed rapidly by 50 ml of 25% sodium thiosulfate intravenously in an adult, or 1.65 ml/kg in children. In certain circumstances, for example, when the diagnosis is uncertain, administration of only the sodium thiosulfate component of the antidote kit may be appropriate. Outside of the United States, hydroxocobalamin, an alternative antidote, may be available.

References

Holstege CP, Isom G, Kirk MA. Cyanide and hydrogen sulfide. In: Flomenbaum NE, Goldfrank LR, Hoffman RS, Howland MA, Lewin NA, Nelson LS. *Goldfrank's Toxicologic Emergencies, 8th Edition*. McGraw-Hill. New York, NY. 2006. p. 1716–1724.

Howland MA. Nitrites. In: Flomenbaum NE, Goldfrank LR, Hoffman RS, Howland MA, Lewin NA, Nelson LS. *Goldfrank's Toxicologic Emergencies, 8th Edition*. McGraw-Hill. New York, NY. 2006. p. 1725–1727.

Howland MA. Sodium thiosulfate. In: Flomenbaum NE, Goldfrank LR, Hoffman RS, Howland MA, Lewin NA, Nelson LS. *Goldfrank's Toxicologic Emergencies, 8th Edition*. McGraw-Hill. New York, NY. 2006. p. 1728–1730.

Vetter J. Plant cyanogenic glycosides. Toxicon 2000;38:11–36.

Poisoning by Plants with Gastrointestinal Toxins

Many and various plant genera are associated with this syndrome.

Toxic Mechanism

Several different mechanisms are utilized by plant toxin to produce gastrointestinal effects, generally described as either mechanical irritation or a pharmacologic effect. Irritant toxins indirectly stimulate contraction of the gastrointestinal smooth muscle. The pharmacologically active agents most commonly work by stimulation of cholinergic receptors in the gastrointestinal tract to induce smooth muscle contraction [e.g., cholinergic (including nicotine-like)] alkaloids. Some plant toxins (e.g., mitotic inhibitors, toxalbumins) alter the normal development and turnover of gastrointestinal lining cells and induce sloughing of this cellular layer. Hepatotoxins may directly injure the liver cells, commonly through the production of oxidant metabolites. Indirect hepatotoxicity may occur, as with the pyrrolizidine alkaloids (see "Poisoning by Plants with Pyrrolizidine Alkaloids", p 31).

Clinical Manifestations

Nausea, vomiting, abdominal cramping, and diarrhea are the hallmarks. Vomiting may be bloody or may contain acid-degraded blood ("coffee grounds") leaked secondary to gastric irritation. Extensive diarrhea and vomiting may produce acid–base, electrolyte, and fluid abnormalities, leading to hypokalemia and profound volume depletion. Small children in particular may become rapidly volume depleted and it may be more difficult to diagnose than in adults. Certain plant toxins that produce prominent gastrointestinal findings may subsequently produce systemic toxicity following absorption. For agents in this group (e.g., mitotic inhibitors, toxalbumins), the gastrointestinal manifestations serve as a warning for potential systemic toxicity.

Specific Therapeutics

Vomiting may be mitigated by antiemetic agents such as metoclopramide; occasionally, resistant emesis may require a serotonin antagonist such as ondansetron. Specific treatment of a patient's diarrhea (e.g., loperamide) is generally unnecessary. Assessment for and correction of volume depletion and metabolic changes are critical. For most patients, intravenous rehydration should be initiated using normal saline or lactated Ringer's solution and adjusted based on clinical or laboratory criteria. Oral rehydration therapy may be attempted in patients with minor clinical abnormalities. Electrolyte and acid–base derangements usually resolve with supportive care but may occasionally require specific therapy. Pharmacotherapies for the prevention of treatment of hepatotoxicity are varied, but empiric therapy with N-acetylcysteine is often suggested.

References

Palmer M, Betz JM. Plants. In: Flomenbaum NE, Goldfrank LR, Hoffman RS, Howland MA, Lewin NA, Nelson LS. *Goldfrank's Toxicologic Emergencies, 8th Edition.* McGraw-Hill. New York, NY. 2006. p. 1577–1602.

Poisoning by Plants with Mitotic Inhibitors

Examples of plant genera associated with this syndrome:

Bulbocodium	*Catharanthus*	*Colchicum*
Gloriosa	*Podophyllum*	

Toxic Mechanism

These agents interfere with the polymerization of microtubules, which must polymerize for mitosis to occur, leading to metaphase arrest. Rapidly dividing cells (e.g., gastrointestinal or bone marrow cells) typically are affected earlier and to a greater extent than those cells that divide slowly. In addition, microtubules are important in the maintenance of proper neuronal function.

Clinical Manifestations

Patients typically have early gastrointestinal abnormalities, including vomiting and diarrhea. Oral ulcers and frank gastrointestinal necrosis can occur. Multisystem organ failure may follow. Bone marrow toxicity typically manifests as an initial leukocytosis, due to release of stored white blood cells, followed by leukopenia. Death may occur from direct cellular toxic effects or from sepsis. Nervous system toxicity, including ataxia, headache, seizures, and encephalopathy, may develop initially, and peripheral neuropathy may develop in patients who survive.

Specific Therapeutics

Initial management includes aggressive supportive and symptomatic care. In patients with profound bone marrow toxicity, colony-stimulating factors may be beneficial. Consultation with appropriate specialists, such as a hematologist, should be strongly considered.

References

Mullins ME, Carrico EA, Horowitz BZ. Fatal cardiovascular collapse following acute colchicine ingestion. J Toxicol Clin Toxicol 2000;38:51–54.

Schier J. Colchicine and podophylline. In: Flomenbaum NE, Goldfrank LR, Hoffman RS, Howland MA, Lewin NA, Nelson LS. *Goldfrank's Toxicologic Emergencies, 8th Edition.* McGraw-Hill. New York, NY. 2006. p. 580–589.

Poisoning by Plants with Nicotine-Like Alkaloids

Examples of plant genera associated with this syndrome:

Baptisia	*Caulophyllum*	*Conium*
Gymnocladus	*Hippobroma*	*Laburnum*
Lobelia	*Nicotiana*	*Sophora*

Toxic Mechanism

These agents are direct-acting agonists at the nicotinic subtype of the acetylcholine receptor in the ganglia of both the parasympathetic and sympathetic limbs of the autonomic nervous system (N_N receptors), the neuromuscular junction (N_M receptors), and the brain.

Clinical Manifestations

Sympathetic stimulation, including hypertension, tachycardia, and diaphoresis, and parasympathetic stimulation, including salivation and vomiting, are common (N_N). Hyperstimulation at the N_M results in fasciculations, muscular weakness, and, rarely, depolarizing neuromuscular blockade. Seizures may occur as a result of effects at cerebral nicotinic receptors.

Specific Therapeutics

Control of the patient's autonomic hyperactivity is generally not needed unless secondary complications, such as myocardial ischemia, develop or are anticipated. In this case, the vital sign abnormalities may be corrected through the judicious use of antihypertensive drugs, including nitroprusside or diltiazem, as appropriate. Neuromuscular symptoms cannot be effectively antagonized because effective agents (e.g., curare-like drugs) would also produce neuromuscular blockade. Patients with inadequate ventilatory effort should be managed supportively. Seizures should respond to intravenous benzodiazepine, such as lorazepam or diazepam.

References

Palmer M, Betz JM. Plants. In: Flomenbaum NE, Goldfrank LR, Hoffman RS, Howland MA, Lewin NA, Nelson LS. *Goldfrank's Toxicologic Emergencies, 8th Edition.* McGraw-Hill. New York, NY. 2006. p. 1577–1602.

Rogers AJ, Denk LD, Wax PM. Catastrophic brain injury after nicotine insecticide ingestion. J Emerg Med 2004;26:169–172.

Solomon ME. Nicotine and tobacco preparations. In: Flomenbaum NE, Goldfrank LR, Hoffman RS, Howland MA, Lewin NA, Nelson LS. *Goldfrank's Toxicologic Emergencies, 8th Edition.* McGraw-Hill. New York, NY. 2006. p. 1221–1230.

Vetter J. Poison hemlock (*Conium maculatum* L.). Food Chem Toxicol 2004;42: 1373–82.

Poisoning by Plants with Pyrrolizidine Alkaloids

Examples of plant genera associated with this syndrome:

Crotalaria	*Echium*	*Heliotropium*
Senecio	*Sesbania*	

Toxic Mechanism

Pyrrolizidine alkaloids are metabolized to pyrroles, which are alkylating agents that injure the endothelium of the hepatic sinusoids or pulmonary vasculature. Endothelial repair and hypertrophy result in venoocclusive disease. Centrilobular necrosis may occur following acute, high-dose exposures, presumably caused by the overwhelming production of the pyrrole. Chronic use is also associated with hepatic carcinoma.

Clinical Manifestations

Acute hepatotoxicity caused by massive pyrrolizidine alkaloid exposure produces gastrointestinal symptoms, right upper quadrant abdominal pain, hepatosplenomegaly, and jaundice as well as biochemical abnormalities consistent with hepatic necrosis [e.g., aspartate ammotransferase (AST), bilirubin, increased international normalized ratio (INR)]. Prolonged, lower-level exposure produces more indolent disease, and patients may present with cirrhosis or ascites caused by hepatic venous occlusion. This syndrome is clinically and pathologically similar to the Budd–Chiari syndrome.

Certain pyrrolizidine alkaloids (e.g., that from *Crotalaria spectabilis*) produce pulmonary vasculature occlusion and the syndrome of pulmonary hypertension in animals, but it is not known whether there is an analogous human response.

Specific Therapeutics

Standard supportive care may allow for some spontaneous repair. There are no known specific therapies. Liver transplantation may be an option for patients with severe hepatotoxicity or cirrhosis.

References

Palmer M, Betz JM. Plants. In: Flomenbaum NE, Goldfrank LR, Hoffman RS, Howland MA, Lewin NA, Nelson LS. *Goldfrank's Toxicologic Emergencies, 8ᵗʰ Edition.* McGraw-Hill. New York, NY. 2006. p. 1577–1602.

Stewart MJ, Steenkamp V. Pyrrolizidine poisoning: A neglected area in human toxicology. Therapeutic Drug Monitoring 2001;23:698–708.

Poisoning by Plants with Sodium Channel Activators

Examples of plant genera associated with this syndrome:

Aconitum	*Kalmia*	*Leucothoe*
Lyonia	*Pernettya*	*Pieris*
Rhododendron	*Schoenocaulon*	*Veratrum*
Zigadenus		

Toxic Mechanism

These agents stabilize the open form of the voltage-dependent sodium channel in excitable membranes, such as neurons and the cardiac conducting system. This causes persistent sodium influx (i.e., persistent depolarization) and prevents adequate repolarization leading to seizures and dysrhythmias, respectively. In the heart, the excess sodium influx activates calcium exchange, and the intracellular hypercalcemia increases both inotropy and the potential for dysrhythmias.

Clinical Manifestations

Vomiting is very common and occurs through a central nervous system-mediated mechanism. Sodium channel effects on sensory neurons may produce paresthesias in a perioral and distal extremity distribution. Persistent depolarization of motor neurons produces fasciculations, motor weakness, and ultimately paralysis. In the heart, the effects of sodium channel opening have been compared to that of the cardioactive steroids: atropine-sensitive sinus bradycardia, atrioventricular blocks, repolarization abnormalities, and, occasionally, ventricular dysrhythmias. However, although the clinical findings are similar, the underlying mechanisms and treatments may differ.

Specific Therapeutics

Normal saline should be rapidly infused into patients with hypotension, and atropine is often therapeutic for sinus bradycardia and conduction blocks. Hypotension may require pressor agents such as norepinephrine. Mechanism-based therapy suggests the use of sodium channel blocking drugs such as lidocaine or amiodarone. None has been proven superior, and the agent used should probably be based on the comfort level of the provider. Although the clinical presentation is similar to poisoning by cardioactive steroids, there is no defined role for digoxin-specific Fab.

References

Lewin NA, Nelson LS. Antidysrhythmics. In: Flomenbaum NE, Goldfrank LR, Hoffman RS, Howland MA, Lewin NA, Nelson LS. *Goldfrank's Toxicologic Emergencies, 8th Edition.* McGraw-Hill. New York, NY. 2006. p. 959–971.

Lin CC, Chan TY, Deng JF. Clinical features and management of herb-induced aconitine poisoning. Ann Emerg Med 2004;43:574–579.

Palmer M, Betz JM. Plants. In: Flomenbaum NE, Goldfrank LR, Hoffman RS, Howland MA, Lewin NA, Nelson LS. *Goldfrank's Toxicologic Emergencies, 8th Edition.* McGraw-Hill. New York, NY. 2006. p. 1577–1602.

Poisoning by Plants with Toxalbumins

Examples of plant genera associated with this syndrome:

Abrus	*Hura*	*Jatropha*
Momordica	*Phoradendron*	*Ricinus*
Robinia		

Toxic Mechanism

The protein toxins derived from these plants work specifically by inhibiting the function of ribosomes, the subcellular organelle responsible for protein synthesis. The toxins typically have two linked polypeptide chains. One of the chains binds to cell-surface glycoproteins to allow endocytosis into the cell. The other chain upon cell entry binds the 60S ribosomal subunit and impairs its ability to synthesize protein.

Clinical Manifestations

Clinical manifestations depend largely on the route of exposure. Following ingestion, local gastroenteritis produces diarrhea and abdominal pain. Because the seed coat of many toxalbumin containing seeds is tough, chewing is generally a prerequisite for toxicity. Absorption of toxin into the systemic circulation allows widespread distribution and multisystem organ failure. Parenteral administration via injection similarly produces diffuse organ dysfunction. Following inhalation of aerosolized toxin, localized pulmonary effects are of greatest concern, although systemic toxicity is possible. Depending on the dose and route administered, the development of findings may be delayed.

Specific Therapeutics

Death from multisystem organ failure is best prevented through aggressive support of vital organ function and prevention of infection. Work is progressing on the use of antiricin antibodies, but it is not in current clinical use.

References

Audi J, Belson M, Patel M, Schier J, Osterloh J. Ricin poisoning: A comprehensive review. JAMA 2005;294:2342–2351.

Bradberry SM, Dickers KJ, Rice P, Griffiths GD, Vale JA. Ricin poisoning. Toxicol Rev 2003; 22:65–70.

SECTION 3.

Plant-Induced Dermatitis (Phytodermatitis)

Skin-associated complaints are the most common form of plant poisoning reported to Poison Control Centers. Dermatologists often see patients with complaints directly or indirectly related to plant exposures (Table 4). For example, outdoor workers may directly develop dermatitis from the toxin-laden pollen from the various plants in the family Compositae (Asteraceae) (e.g., chrysanthemums, which contain sesquiterpene lactones). Indirectly, the use of perfume or other lotions that contain plant derivatives may produce dermatitis in an unsuspecting user.

Plant-induced dermal disorders are frequently categorized into several groups to generalize their clinical effects and management, but plant-specific mechanisms and therapies may exist:

1. Mechanical irritants
2. Chemical irritants
3. Allergens
4. Phototoxins
5. Pseudophytodermatitides

The majority of these complaints are managed by primary care physicians or dermatologists. This section provides a brief overview of the problem and should be supplemented by the use of a reference text or the advice of a dermatologist. Poison Control Centers, physicians, botanists, and toxicologists are frequently confronted by situations in which a plant is implicated in the formation of a rash, suggesting that the ability to recognize plant-induced dermatitis is important for all.

Mechanical Irritants

Mechanical injury, sometimes called toxin-mediated urticaria, is generally induced by plants with obvious physical characteristics that directly injure the skin, such as the barbs of aloe or the trichomes of stinging nettles (*Urtica dioica*) (Table 5). In the latter case, the stingers are fragile hypodermic syringe-like tubules that contain a mixture of irritant chemicals which are injected into the skin after the trichome breaks the dermal barrier. Following exposure, patients rapidly develop short-lived wheals with intense pruritis. Less-obvious exposures occur when the idioblasts found in common houseplants such as *Dieffenbachia*

TABLE 4. Representative Plants Causing Contact Dermatitis

Botanical name	Common name
Amaryllidaceae	
Narcissus species	Daffodil, Narcissus, Jonquil
Anacardiaceae	
See Table 8	
Annonaceae	
Asimina triloba (L.) Dunal	Pawpaw
Apocynaceae	
Allamanda cathartica L.	Allamanda, Canario
Nerium oleander L.	Oleander
Araceae	
See Table 6	
Araliaceae	
Hedera canariensis Willd.	Algerian Ivy
H. helix L.	English Ivy
Aristolochiaceae	
Aristolochia elegans M.T. Mast	Calico Flower
A. gigantea Mart. & Zucc. non Hook.	
A. grandiflora Swartz	Pelican Flower
Asclepiadaceae	
See Table 7	
Bignoniaceae	
Campsis radicans (L.) Seem.	Trumpet Creeper
Bromeliaceae	
Ananas comosus (L.) Merrill	Pineapple
Chenopodiaceae	
Sarcobatus vermiculatus (Hook.) Torr.	Greasewood
Commelinaceae	
Rhoeo spathacea (Swartz) Stearn (= *R. discolor* (L'Her.) Hance)	Moses-in-a-Boat, Oyster Plant
Setcreasea pallida Rose cv. 'Purple Heart' (= *S. purpurea* Boom)	Purple Queen
Compositae (Asteraceae)	
Ambrosia species	Ragweed
Artemisia species	Mugwort
Aster species	Aster, Daisy
Chrysanthemum species	Chrysanthemum, Daisy, Feverfew, Marguerite
Erigeron species	Fleabane
Franseria acanthicarpa (Hook.) Coville (= *Ambrosia acanthicarpa* Hook.)	Poverty Weed
Gaillardia species	Gaillardia
Helenium autumnale L.	Sneezeweed
H. microcephalum DC.	
Iva species	Marsh Elder
Lactuca sativa L.	Lettuce
Oxytenia acerosa Nutt. (= *Iva acerosa* (Nutt.) R.C. Jackson)	Copper Weed

TABLE 4. Representative Plants Causing Contact Dermatitis, *Continued*

Botanical name	Common name
Parthenium argentatum Gray	Guayule
P. hysterophorus L.	Parthenium
Rudbeckia hirta L. (=*R. serotina* Nutt.)	Black-eyed Susan
Soliva pterosperma (Juss.) Less.	Spurweed
Tanacetum vulgare L.	Tansy
Tagetes minuta L.	Mexican Marigold
Xanthium species	Cocklebur
See Table 9	
Convolvulaceae	
Dichondra repens J.R. & G. Forst.	
Cornaceae	
See Table 5	
Euphorbiaceae	
Hura crepitans L.	Monkey Pistol, Sandbox Tree, Javillo
Ricinus communis L.	Castor Bean, Higuereta, Ricino
See Table 7	
Fumariaceae	
Dicentra spectabilis (L.) Lem.	Bleeding Heart
Ginkgoaceae	
Ginkgo biloba L.	Ginkgo, Maidenhair Tree
Gramineae (Poaceae)	
Oryza sativa L.	Rice
Panicum glutinosum Sw.	Sticky Grass
Secale cereale L.	Rye
Hydrophyllaceae	
Phacelia campanularia Gray	California Bluebell
P. crenulata Torr. ex S. Wats.	Scorpion Flower, Scorpion Weed
P. minor (Harv.) Thell. ex F. Zimm. (= *P. whitlavia* A. Gray; *Whitlavia grandiflora* Harv.)	Whitlavia
P. parryi Torr.	
P. viscida (Benth. ex Lindl.) Torr.	
See Table 5	
Juglandaceae	
Juglans nigra L.	Black Walnut
Leguminosae (Fabaceae)	
Prosopis glandulosa Torr. (= *P. juliflora* (Sw.) DC. var *glandulosa* (Torr.) Cockerell)	Mesquite
See Table 5	
Liliaceae	
Allium cepa L.	Onion, Cebolla
A. sativum L.	Garlic, Ajo
Hyacinthus species	Hyacinth
Tulipa species	Tulip

TABLE 4. Representative Plants Causing Contact Dermatitis, *Continued*

Botanical name	Common name
Loranthaceae	
Phoradendron serotinum (Raf.) M.C. Johnst. (= *P. flavescens* (Pursh) Nutt.)	(American) Mistletoe
Magnoliaceae	
Magnolia grandiflora L.	Magnolia, Bull Bay
Moraceae	
Maclura pomifera (Raf.) C.K. Schneid.	Osage Orange
See Table 9	
Myrtaceae	
Eucalyptus globulus Labill.	Blue Gum, Eucalyptus
Orchidaceae	
Cypripedium species	Lady's Slipper
Palmae (Arecaceae)	
See Table 6	
Pinaceae	
Abies balsamea (L.) Mill.	Balsam Fir
Primulaceae	
Primula farinosa L.	Birdseye Primrose
P. obconica Hance	Primula, German Primrose
Proteaceae	
Grevillea banksii R. Br.	Kahili Flower
G. robusta A. Cunn.	Silk(y) Oak
Ranunculaceae	
See Table 7	
Rosaceae	
Agrimonia species	Agrimony
Rosa odorata (Andr.) Sweet	Tea Rose
See Table 9	
Rutaceae	
See Table 9	
Saxifragaceae	
Hydrangea species	Hydrangea
Solanaceae	
Lycopersicon esculentum Mill.	Tomato
Solanum carolinense L.	Horse Nettle
S. tuberosum L.	Potato
Thymelaeaceae	
See Table 7	
Ulmaceae	
Ulmus glabra Huds.	Wych Elm, Scotch Elm
U. procera Salisb.	English Elm
Umbelliferae (Apiaceae)	
See Table 9	
Urticaceae	
See Table 5	

TABLE 4. Representative Plants Causing Contact Dermatitis, *Continued*

Botanical name	Common name
Vitaceae	
See Table 6	
Zygophyllaceae	Creosote Bush
Larrea tridentata (Sesse & Moc. ex DC.)	
Coville (= *L. glutinosa* Englem.)	

TABLE 5. Representative Plants with External Irritant, Stinging Hairs, or Detachable Needles

Botanical name	Common name
Cactaceae	
Opuntia species (e.g., *O. microdasys* (Lehm.) Pfeiff.)	Bunny Ears, Prickly Pear
Cannabaceae	
Humulus lupulus L.	Hops
Cornaceae	
Cornus sanguinea L.	Bloodtwig Dogwood
Euphorbiaceae	
Acidoton urens Sw.	Mountain Cowitch
Cnidoscolus chayamansa McVaugh	Chaya
C. stimulosus (Michx.) Engelm.	Tread Softly, Bull Nettle
C. texanus (Muell. Arg.) Small	
C. urens (L.) Arth. (= *Jatropha urens* L.)	
Dalechampia scandens L.	Liane Gratte
Platygyne hexandra (Jacq.) Muell. Arg.	Pringamosa
Tragia volubilis L.	Pringamosa
Hydrophyllaceae	
Phacelia imbricata Greene	
P. malvifolia Cham.	Stinging Phacelia
Wigandia caracasana H.B.K.	
W. urens (Ruiz & Pav.) H.B.K.	
Leguminosae (Fabaceae)	
Lupinus hirsutissimus Benth.	Stinging Lupine
Mucuna deeringiana (Bort) Merrill	Velvet Bean
(= *Stizlobium deeringianum* Bort)	Cowhage, Cowitch, Pica-Pica,
M. pruriens DC. (= *Stizlobium pruriens*	Pois Gratté, Vine Gungo Pea
(L.) Medik.; *Dolichos pruriens* L.)	
M. urens (L.) DC.	Bejuco Jairey, Ox-Eye Bean,
	Yeaux Bourrique, Torteza
Malpighiaceae	
Malpighia polytricha A. Juss.	Touch-Me-Not
M. urens L.	Cowitch Cherry
Sterculiaceae	
Sterculia apetala (Jacq.) Karst. (in exposed fruit)	
Urticaceae	
Laportea aestuans (L.) Chew (= *Fleurya aestuans* (L.) Gaud.)	Pica-Pica
L. canadensis (L.) Weddell	Wood Nettle
Urera baccifera (L.) Weddell	Ortiga Brava
Urtica dioica L.	Stinging Nettle
U. urens L.	Stinging Nettle

spp. injure the skin. These idioblasts contain both needle-like calcium oxalate crystals (mechanical irritants) in a soup of irritant chemicals (chemical irritants; Table 6). The crystals are forcibly injected into the skin or mucosa following mechanical stimulation of the idioblasts. Depending on the anatomical location of the crystal deposition, the clinical effects can be minor (e.g., skin), consequential (e.g., eye), or rarely lethal (e.g., airway). Treatment is generally supportive and symptomatic in nature. Skin involvement with a mechanical irritant should be treated with demulcent cremes, ice, and analgesics, and perhaps removal of the offending agent if appropriate. Ocular involvement is similarly managed with symptomatic care that may include ocular irrigation and systemic analgesics, and most exposures should prompt consultation with an ophthalmologist. Oropharyngeal exposures mandate rigorous attention to the airway, and patients may require corticosteroids to limit pharyngeal swelling. Endoscopy is likely indicated in any patient with clinical findings consistent with airway involvement (e.g., dysphonia, dysphagia, stridor).

Chemical Irritants

Chemical irritants differ from the mechanical irritants in that they produce their clinical effects on the basis of a physicochemical quality of the toxin rather than through overt mechanical means (Table 7). Some of these toxins may be introduced along with mechanical irritants, as already noted. Chemical irritants may be directly irritating on the basis of pH or other chemical effects (e.g., solubility), mimicry of an endogenous compound (e.g., acetylcholine or histamine), enzymatic damage, or activation of inflammation. These agents may be protoxins that require metabolic transformation to produce the ultimate toxin [e.g., sinigrin in Brassicaceae (mustard) plants forms allyl isothiocyanate]. Chili peppers (*Capsicum* spp.) contain capsaicin, which induces the release of stored neurotransmitter from sensory neurons (substance P) and produces the deep aching pain characteristic of the "Hunan hand" syndrome, for example. Treatment of chemical-induced irritation includes decontamination by thorough washing of the affected area, analgesics, and symptomatic care.

Allergens

Although any type of allergic response may occur, a type IV, or delayed, hypersensitivity response, also known as allergic contact dermatitis (ACD), is by far the most common caused by plants (Table 8). Many occupations, such as florists, gardeners, and even outdoor workers uninvolved with plant work, are at risk for the development of ACD. Among florists, exposure to Peruvian lily,

TABLE 6. Representative Plants Containing Irritant Raphides

Botanical name	Common name
Araceae	
Alocasia species (e.g., *A. macrorrhiza* (L.) G. Don)	Elephant's Ear, Taro
Anthurium andreanum Linden	Flamingo Lily
Arum italicum Mill.	Italian Arum
A. maculatum L.	Cuckoopint
Caladium bicolor (Ait.) Venten.	Caladium
Calla palustris L.	Water Arum, Marsh Marigold
Colocasia species (e.g., *C. esculenta* (L.) Schott	Elephant's Ear
Dieffenbachia species	Dumbcane
Epipremnum aureum (Linden & André)	Pothos
Bunt. (= *Pothos aureus* Linden & André;	
Raphidophora aurea (Linden & André)	
Birdsey; *Scindapsus aureus* (Linden & André) Engl. &	
K. Krause)	
Philodendron scandens C. Koch & H. Sello ssp. *oxycardium*	Heartleaf Philodendron
(Schott) Bunt.	
P. selloum C. Koch	
Palmae (Arecaceae)	
Caryota mitis Lour.	Fishtail Palm
Vitaceae	
Parthenocissus quinquefolia (L.) Planch.	Virginia Creeper
P. triscuspidata (Siebold & Zucc.) Planch.	Boston Ivy

the very common centerpiece flower, results in ACD because of the common sensitizer tulipin A. The general pathogenesis of this reaction involves a primary exposure to a toxin resulting in an immune response (i.e., sensitization) developing in the affected individual. In some cases, particularly with toxins that are too small to elicit an immune response, the binding of the toxin or its metabolite to an endogenous compound (i.e., as a hapten) results in immunological recognition. Upon reexposure to the same, or closely related, toxin the primed immune system recognizes the antigen (or haptenized endogenous compound), and an immunological response is triggered. The result is a slowly developing (over hours to days) rash, consisting typically of pain, itch, redness, swelling, and blisters localized to the affected area. The sensitizing potential of the various plant-borne toxins varies, but urushiol is among the most potent, and "poison ivy" is among the most frequently encountered sensitizers. Nearly everyone is capable of being sensitized to urushiol, accounting for the reason that "Rhus dermatitis" is given its own moniker. Although the reaction resembles irritant dermatitis, it is more slow to develop and requires previous exposure. Many

TABLE 7. Representative Plants Containing an Irritant Sap or Latex

Botanical name	Common name
Agavaceae	
Agave species (e.g., *A. americana* L.)	Century Plant, Maguey
Apocynaceae	
Acokanthera oblongifolia (Hochst.) Codd (= *A. spectabilis* (Sond.) Hook. f.)	Bushman's Poison, Wintersweet
Plumeria species	Frangipani
Asclepiadaceae	
Calotropis gigantea (L.) Ait. f.	Crown Flower
C. procera (Ait.) Ait. f.	Algodón de Seda
Euphorbiaceae	
Euphorbia cotinifolia L.	Poison Spurge, Carrasco
E. gymnonota Urb.	
E. lactea Haw.	Candelabra Cactus
E. lathyris L.	Caper Spurge, Mole Plant
E. marginata Pursh	Snow-on-the-Mountain
E. milii Ch. des Moulins	Crown-of-Thorns
E. myrsinites L.	Creeping Spurge
E. tirucalli L.	Pencil Cactus
Excoecaria agallocha L. var. *orthostichalus* Muell. Arg.	Blinding Tree
Grimmeodendron eglandulosum (A. Rich.) Urb.	Poison Bush
Hippomane mancinella L.	Beach Apple, Manzanillo
Pedilanthus tithymaloides (L.) Poit.	Slipper Flower
Sapium hippomane G.F.W. Mey.	
S. laurocerasus Desf.	Hinchahuevos
Stillingia sylvatica Gard.	Queen's Delight
Synadenium grantii Hook. f.	African Milkbush
Ranunculaceae	
Caltha palustris L.	Pasque Flower
Clematis species (e.g., *C. virginiana* L.) *Pulsatilla patens* Mill. (= *Anemone patens* L.)	Marsh Marigold Virgin's Bower
Ranunculus species (e.g., *R. acris* L.)	Buttercup, Crowfoot
Thymelaeaceae	
Daphne mezereum L.	Daphne
Dirca palustris L.	Leatherwood, Wicopy

related urushiol-like compounds from diverse sources [e.g., from mango (*Mangifera indica*) or cashew nut (*Anacardium occidentale*)] produce identical reactions in patients sensitized to urushiol.

Diagnosis includes the use of patch testing, in which a single or several known allergens are applied to the skin and a reaction is sought and is

TABLE 8. Anacardiaceae Producing Allergic Contact Dermatitis

Botanical name	Common name
Anacardium occidentale L.	Cashew, Marañón
Comocladia species (e.g., *C. dodonaea* (L.) Urban)	Guao
Cotinus coggygria Scop. (= *Rhus cotinus* L.)	Smoke Tree
Mangifera indica L.	Mango
Metopium toxiferum (L.) Krug & Urban	Poisonwood, Cedro Prieto
Schinus terebinthifolius Raddi	Brazilian Pepper Tree, Florida Holly
Toxicodendron diversilobum (Torr. & A. Gray) Greene (= *Rhus diversiloba* Torr. & A. Gray)	Western Poison Oak
T. pubescens P. Mill. (= *T. toxicarium* (Salisb.) Gillis; *Rhus toxicodendron* L.; *R. quercifolia* (Michx.) Steudel)	Poison Ivy Western Poison Ivy Eastern Poison Oak
T. radicans (L.) Kuntze (=*Rhus radicans* L.)	
T. rydbergii (Small) Greene	
T. vernix (L.) Kuntze (= *Rhus vernix* L.)	Poison Sumac

generally confirmatory. Although more advanced testing is available, this is a common initial screen for contact allergens. Occasionally, irritant dermatitis may result and be misinterpreted as ACD, so expert interpretation is needed. The risk of patch screening by this method is that sensitization to any of the tested compounds may occur just from the testing alone, so often the most strongly sensitizing plants are excluded from testing.

The primary therapy, of course, is avoidance of the known allergen. Because this is difficult or impossible in some situations, the use of barrier protection may provide a sufficient impediment to dermal exposure. Barriers include the use of clothing or of barrier creams that can be applied if an exposure is anticipated. Treatment of the ACD once it has occurred is generally symptomatic, with the use of analgesics, antipruritic medications (e.g., hydroxyzine, diphenhydramine), occasionally corticosteroids (generally topically applied), and rarely immune modulating agents (e.g., tacrolimus ointment). Desensitization may be attempted for patients with severe reactions or unavoidable exposures.

Phototoxins

This is a relatively uncommon clinical entity, in which certain compounds increase the sensitivity of the skin (photodermatitis) to ultraviolet light (e.g.,

TABLE 9. Representative Plants Producing Phytophotodermatitis

Botanical name	Common name
Compositae (Asteraceae)	
Achillea millefolium L.	Yarrow, Milfoil, Milenrama
Anthemis cotula L.	Dog Fennel, Mayweed
Moraceae	
Ficus carica L.	Fig
F. pumila L.	Creeping Fig, Creeping Rubber Plant
Rosaceae	
Agrimonia eupatoria L.	Agrimony
Rutaceae	
Citrus aurantiifolia (Christm.) Swingle	Lime
Dictamnus albus L.	Gas Plant, Burning Bush
Pelea anisata H. Mann	Mokihana
Ruta graveolens L.	Rue, Ruda
Umbelliferae (Apiaceae)	
Ammi majus L.	Bishop's Weed
Anthriscus sylvestris (L.) Hofmann	
Daucus carota L. var. *carota*	Queen Anne's Lace, Wild Carrot
Daucus carota var. *sativus* Hoffm.	Carrot
Heracleum lanatum Michx.	Cow Parsnip
H. mantegazzianum Sommier & Levier	Giant Hogweed, Wild Rhubarb
H. sphondylium L.	Cow Parsnip
Pastinaca sativa L.	Parsnip

sunlight) (Table 9). Classically, psoralen, a furocoumarin derived from celery and other plants, enters the skin either directly by contact or via the systemic circulation following ingestion. In the skin, the psoralen is activated by sunlight to produce oxidant skin damage, which manifests as burning, erythematous skin in sun-exposed areas, which may blister severely (i.e., sunburn). Interestingly, psoralens may be administered therapeutically to patients with dermal disorders such as psoriasis to increase the sensitivity of the skin to therapeutic ultraviolet light.

Pseudophytodermatitides

Given the ubiquity of plants and the constant interaction of humans with plants, many dermatological disorders are often attributed to a contemporaneous plant exposure. However, mimics of plant dermatitis are common and may be missed due to the often-simultaneous nature of the two exposures. For example, pesticides, fungicides, insects, and soil products may each induce dermatitis that is often indistinguishable from one of the foregoing syndromes.

Without intense investigation or advanced medical testing, this link may be missed and the patient advised incorrectly to avoid a certain plant exposure. More consequentially, the patient may not be aware of the dermatitis trigger. The best method by which to distinguish pseudophytodermatitis from phytodermatitis is by having the necessary clinical suspicion and attentiveness to the exposure. Even then, this link is often difficult to make.

SECTION 4.

Gastrointestinal
Decontamination

Gastrointestinal (GI) decontamination is the use of medical methods to decrease absorption of ingested toxins. Not long ago, it was widely believed that the proper initial approach to the management of all patients with toxic exposures included an aggressive attempt at gastrointestinal decontamination. Its widespread use was based on the assumption that these methods were effective at toxin removal and that they improved patient outcomes. Despite the fact that, for most toxins, until the toxin is absorbed into the body it is not "poisonous," it has been quite difficult to prove that preventing absorption improves outcome.

Over the past decade, with the publication of several studies of the efficacy of GI decontamination for poisonings in general, it has become clear that what was not adequately considered previously was the risk of the decontamination procedure itself. That is, if a type of procedure harms more patients than it helps, then that procedure should not generally be employed. An even more refined approach would take into consideration the individual patient; that is, if a given procedure is more likely to harm rather than help a given patient, then it should not be utilized for that patient. Fortunately, the risk of all gastrointestinal decontamination procedures is quite small, but it is this very fact that had obscured their potential for harm; that is, to find a small risk of harm to a population, very large numbers of patients need to be studied. The vast majority of plant exposures are to nontoxic plants or to nontoxic amounts of potentially toxic plants. However, to optimize outcome, as the risk of harm or death from a poisoning rises, the acceptable risk from a gastrointestinal decontamination becomes larger.

This new concept places an unseen burden on those at the interface of toxicology and medicine, including botanists and physicians, to better understand the risks truly associated with any specific type of plant exposure. Realize that the operative word here is "exposure," because most cases discussed with Poison Control Centers and faced by emergency physicians and other clinicians are not actually poisonings but rather "potential poisonings." That is, it is common, particularly in children, that either the ingestion of the plant did not actually occur or that if it did occur it was in a subtoxic quantity. Thus, the perceived risk of harm from the poison must be tempered by the risk that the exposure was consequential, and both balanced against the risk of harm from the therapy. Because there are not very large numbers of plant-poisoned patients, particularly those poisoned by any single type of plant, determination of the risk of most plants has been elusive.

Described next is a rational approach to the initial management of patients with plant poisonings, specifically gastrointestinal decontamination. The remainder of the book contains details concerning the known and assumed risks associated with exposure to the various plants.

Syrup of Ipecac

There are few circumstances in which the induction of emesis (vomiting) is the preferred method of gastrointestinal decontamination. Depending on the plant ingested and the time since exposure, syrup of ipecac may be able to evacuate a substantial amount of the toxin before its absorption into the systemic circulation. In the interest of preventing iatrogenic harm, no patient who has ingested an irritant plant should receive ipecac, as it is likely that the irritated esophagus and oral cavity are likely to be reinjured with vomiting. The specific circumstance in which ipecac may be useful is when a prolonged transport time to healthcare coupled with a very recent time of ingestion (within 15 minutes) is expected for a patient who has likely ingested a toxic quantity of a poisonous plant. Discussion with a Poison Control Center or knowledgeable physician is probably most appropriate before administration. Ipecac is unlikely to be necessary for plant-poisoned patients once they are in the hospital, although there may be rare exceptions. The availability of syrup of ipecac has been dramatically reduced over the past few years, making it less likely that it will be used in most circumstances.

Orogastric Lavage

Orogastric lavage involves the insertion of a tube through the mouth or nose into the stomach to evacuate the stomach contents ("pump the stomach"). Water is then instilled through the tube and removed several times to "wash" the stomach free of its contents. Depending on the plant ingested and the time since exposure, orogastric lavage may be able to evacuate a substantial amount of the toxin before its absorption into the systemic circulation. However, because orogastric lavage is not without risk of complication (e.g., esophageal injury, pulmonary aspiration), the use of this technique should be limited to patients who have ingested a toxic plant that is likely to carry significant consequences. In general, performance of this technique is limited to trained hospital personnel and should, in most cases, be done in consultation with a Poison Control Center or medical toxicologist.

Activated Charcoal

Certain patients should receive oral activated charcoal. This substance is capable of adsorbing many diverse toxins to its porous surface and can prevent the absorption of toxins into the body. Clearly, the ability of activated charcoal to adsorb the vast majority of plant-derived toxins is poorly studied (and unlikely to be studied in the near future). However, based on current understanding of the adsorption characteristics of activated charcoal and the physicochemical properties of many plant toxins (e.g., small molecular weight, uncharged), there is likely to be some benefit provided to most patients. This, coupled with a low risk of side effects (e.g., pulmonary aspiration) makes oral activated charcoal the method of choice for most poisoned patients who require gastrointestinal decontamination. Oral activated charcoal should not be given to patients who have ingested injurious plant substances, those likely to vomit (particularly if they are likely to lose consciousness or seize), or those who are not presumed to have ingested a consequential quantity of a toxic plant (i.e., as punishment).

Whole Bowel Irrigation

Because many plant-borne toxins are ingested in an impervious seed, it is likely that there will be a delay to the release of the toxic compound. In many of these cases, and perhaps in several other situations, the use of whole-bowel irrigation with polyethylene glycol electrolyte solution may be able to "flush" the entire gastrointestinal tract free of plant material before the toxin release occurs. As with the other gastrointestinal decontamination techniques, there are few or no plant-specific studies of the utility or safety of this technique. However, based on the known efficacy and general safety in other poisoned patients, whole-bowel irrigation should be considered in patients who ingest a significant amount of a plant likely to be slowly released from its vehicle (typically its seed). Oral activated charcoal administered concomitantly is often appropriate.

Summary

Despite the lack of evidence proving a benefit, consideration of the need for a gastrointestinal decontamination procedure is indicated for all poisoned patients. Among the factors that influence the choice of procedure is likelihood of and time since ingestion, quantity and nature of the plant and its toxin, the clinical characteristics of the patient, and the risk–benefit relationship of the gastrointestinal decontamination method. Each case needs to be individually considered. Consultation with a medical toxicologist, Poison Control Center, and/or a botanist is often necessary.

References

American Academy of Clinical Toxicology; European Association of Poisons Centres and Clinical Toxicologists. Position paper: Cathartics. J Toxicol Clin Toxicol 2004;42(2):243–253.

American Academy of Clinical Toxicology; European Association of Poisons Centres and Clinical Toxicologists. Position paper: Whole bowel irrigation. J Toxicol Clin Toxicol 2004; 42(2):843–854.

American Academy of Clinical Toxicology; European Association of Poisons Centres and Clinical Toxicologists. Position paper: Gastric lavage. J Toxicol Clin Toxicol 2004;42(2): 933–943.

American Academy of Clinical Toxicology; European Association of Poisons Centres and Clinical Toxicologists. Position paper: Ipecac syrup. J Toxicol Clin Toxicol 2004;42(2): 133–143.

American Academy of Clinical Toxicology; European Association of Poisons Centres and Clinical Toxicologists. Position paper: Single-dose activated charcoal. Clin Toxicol (Phila) 2005;43(2):61–87.

American Academy of Pediatrics Committee on Injury, Violence, and Poison Prevention. Poison treatment in the home. American Academy of Pediatrics Committee on Injury, Violence, and Poison Prevention. Pediatrics 2003;112(5):1182–1185.

Christophersen ABJ, Hoegberg LCG. Techniques used to prevent gastrointestinal absorption. In: Goldfrank LR, Hoffman RS, Howland MA, Lewin NA, Nelson LS. *Goldfrank's Toxicologic Emergencies, 8th Edition*. McGraw-Hill. Hew York, NY. 2006. p. 109–123.

de Silva HA, Fonseka MM, Pathmeswaran A, et al. Multiple-dose activated charcoal for treatment of yellow oleander poisoning: A single-blind, randomised, placebo-controlled trial. Lancet 2003;361:1935–1938.

Manoguerra AS, Cobaugh DJ. Guidelines for the Management of Poisoning Consensus Panel. Guideline on the use of ipecac syrup in the out-of-hospital management of ingested poisons. Clin Toxicol (Phila) 2005;43(1):1–10.

SECTION 5.

Individual Plants

Abrus precatorius L.
Family: Leguminosae (Fabaceae)

Common Names: Bead Vine, Black Eyed Susan, Coral Bead Plant, Crab's Eyes, Graines d'Église, Indian Licorice, Kolales Halomtano, Jequirity/Jequirty Bean, Jumbee Beads, Jumbi Beeds, Jumbo Beads, Licorice Vine, Liane á Réglisse, Love Bead, Love Pea, Ojo de Pájaro, Ojo de Cangrejo, Peonia, Precatory Bean, Prayer Beads, Pukiawe-Lei, Red Bead Vine, Réglisse, **Rosary Pea**, Semilla de Culebra, Seminole Bead, Weather Plant, Weather Vine, Wild Licorice

Abrus precatorius, close-up of seed pod (above)

Description: The rosary pea is a slender, twining vine with a woody base. It is supported generally on other plants or a fence. The compound leaves have numerous short leaflets, which are sensitive to light and droop at night and on cloudy days. The inconspicuous flowers are a pale reddish-purple. The fruit, a pea-shaped pod about 1.5 inches long, splits open as it dries to reveal three to five hard-coated, brilliant scarlet, pea-sized seeds with a small enamel-black spot at the point of attachment

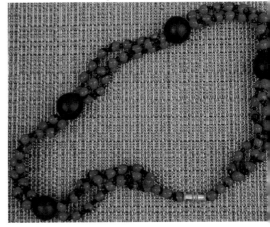

Abrus precatorius, necklace made from seeds (smaller seeds are of this species, larger seeds are from an unidentified species) (below)

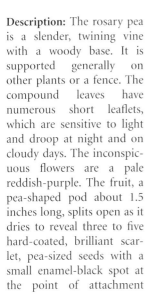

(hilum). This hilar spot serves to distinguish them from look-alike seeds from Mexican vines of the genus *Rhynchosia*, in which the black and red colors are reversed.

Distribution: This weed is common throughout the tropics and subtropics. The seeds are used in jewelry, in handicrafts, and as good-luck charms. Tourists, usually unaware of the toxicity of the seeds, often import them into the United States as keepsakes and gifts.

Toxic Part: The toxin is contained within the hard, water-impermeable coat of the seeds. The toxin is not released unless the seed is chewed and digested or the seed coat is otherwise broken (for example, when the seeds are pierced and threaded on a string as in a necklace).

Toxin: Abrin, a plant lectin (toxalbumin) related to ricin, inhibits cellular protein synthesis and may be extremely toxic.

Clinical Findings: Ingested seeds generally remain intact as they pass through the gastrointestinal tract without releasing any toxin and causing any toxicity. However, if the seeds are chewed, pulverized, or digested (i.e., if passage through the gastrointestinal tract is delayed), then the toxin is absorbed by intestinal cells, causing mild to severe gastrointestinal toxicity. Symptoms depend upon the amount of toxin exposure and include nausea, vomiting, abdominal cramping, diarrhea, and dehydration. Variations in the severity of toxicity may be related to the degree to which the seeds are ground or chewed before ingestion. Parenteral administration (such as by injection or inhalation), or perhaps large ingestion, may produce life-threatening systemic findings, including multisystem organ failure, even with small exposures.

Management: Ingestion of intact seeds does not cause toxicity in the majority of cases and requires no therapy. Cases associated with gastrointestinal symptomatology need to be assessed for signs of dehydration and electrolyte abnormalities. Activated charcoal should be administered. Intravenous hydration, antiemetics, and electrolyte replacement may be necessary in severe cases, particularly those involving children. Consultation with a Poison Control Center should be considered. See "Poisoning by Plants with Toxalbumins," p. 33.

References

Benson S, Olsnes S, Pihl A. On the mechanism of protein-synthesis inhibition by abrin and ricin. Eur J Bioochem 1975;58:573–580.

Davies JH. *Abrus precatorius* (rosary pea): The most common lethal plant poison. J Fla Med Assoc 1978;65:188–191.

Dickers KJ, Bradberry SM, Rice P, Griffiths GD, Vale JA. Abrin poisoning. Toxicol Rev 2003;22:137–142.

Fernando C. Poisoning due to *Abrus precatorius* (jequirity bean). Anaesthesia 2001;56:1178–1180.

Hart M. Hazards to health: jequirty-bean poisoning. N Engl J Med 1963;269:885–886.

Kinamore PA. *Abrus* and *Ricinus* ingestion: Management of three cases. Clin Toxicol 1980; 17:401–405.

Olsnes S. The history of ricin, abrin and related toxins. Toxicol 2004;44:361–370.

Acokanthera species
Family: Apocynaceae
Acokanthera longiflora Stapf
Acokanthera oblongifolia (Hochst.) Codd (=*A. spectabilis* (Sond.) Hook. f.)
Acokanthera oppositifolia (Lam.) Codd
Acokanthera oblongifolia

Common Names: Bushman's Poison, Poison Bush, Poison Tree; **Wintersweet** (A. oblongifolia)

Description: These dense evergreen shrubs or small trees are native to Africa. The leaves are large, leathery, and have smooth edges. The fragrant flowers are tubular with a flat flare at the mouth and are borne in clusters. The fruit resembles a small, ellipsoidal plum and turns reddish to purple-black at maturity; it

Acokanthera oppositifolia (left)

Acokanthera oppositifolia, close-up of branch showing opposite leaves (below)

contains one to two seeds. Injury to the plant causes exudation of copious amounts of white latex, as is characteristic of this plant family.

Distribution: These plants are grown fairly commonly, often as hedges in California, to a much lesser extent in Florida and Hawaii, and as a greenhouse plant elsewhere.

Toxic Part: The fruit pulp contains only traces of toxin and, in some species, is considered edible. The seeds contain the greatest concentration, but the toxin is distributed in substantial amounts throughout the plant, including the wood.

Toxin: Cardioactive steroids resembling digitalis.

Clinical Findings: Poisoning may cause clinical findings typical of cardioactive steroids. Toxicity has a variable latent period that depends on the quantity ingested. Dysrhythmias are usually expressed as sinus bradycardia, premature ventricular contractions, atrioventricular conduction defects, or ventricular tachydysrhythmias. Hyperkalemia, if present, may be an indicator of toxicity.

Management: Gastrointestinal decontamination as appropriate, serial electrocardiograms, and serum potassium determinations should be performed. If serious cardioactive steroid toxicity is considered, digoxin-specific Fab should be administered. Consultation with a Poison Control Center should be considered. See "Poisoning by Plants with Cardioactive Steroids/Cardiac Glycosides," p. 24.

References

Kingston DG, Reichstein T. Cytotoxic cardenolides from *Acokanthera longiflora* Stapf. and related species. J Pharm Sci 1974;63:462–464.

Menziani E, Sancin P, Pertusato G. On the determination of ouabain in *Acokanthera abyssinica* seeds. Boll Chim Farm 1964;103:825–828.

Omino EA, Kokwaro JO. Ethnobotany of Apocynaceae species in Kenya. J Ethnopharmacol 1993;40:167–180.

Radford DJ, Gillies AD, Hinds JA, Duffy P. Naturally occurring cardiac glycosides. Med J Aust 1986;144:540–544.

Aconitum species
Family: Ranunculaceae
Aconitum columbianum Nutt.
Aconitum napellus L.
Aconitum reclinatum Gray
Aconitum uncinatum L.

Common Names: Aconite, Friar's Cap, Helmet Flower, **Monkshood**, Soldier's Cap, Trailing Monkshood, Wild Monkshood, Wolfsbane

Description: These perennial plants are usually erect, sometimes branched, 2 to 6 feet in height, and have tuberous roots. They resemble delphiniums. The characteristic helmet-shaped flowers grow in a raceme at the top of the stalk and appear in summer or autumn. The flowers are usually blue but may be white, pink, or flesh toned. The dried seedpods contain numerous tiny seeds. *Aconitum napellus* is the commonly cultivated monkshood.

Distribution: North America, Eurasia. *Aconitum columbianum* grows in montaine meadows and open woods in the Western United States, north to British Columba and Alaska. *Aconitum napellus* is cultivated and naturalized in New York, Ontario, and Newfoundland. *Aconitum reclinatum* is found in woods and marshes from Pennsylvania to West Virginia, Virginia, North Carolina, and Georgia. *Aconitum uncinatum* is found in woods from Pennsylvania to Ohio, Indiana, North Carolina, Georgia, and Tennessee.

Toxic Part: The whole plant is poisonous, especially the leaves and roots.

Toxin: Aconitine and related alkaloids, sodium channel activators.

Clinical Findings: Exposures are relatively uncommon. However, these plants are utilized in some herbal products (e.g., chuanwu, caowu, fuzi). Symptoms are predominantly neurological and cardiac. There is transient burning in the mouth after ingestion, followed after several hours by increased salivation, vomiting, diarrhea, and a tingling sensation in the skin (paresthesia). The patient may complain of headache, muscular weakness, and dimness of vision. Bradycardia and other cardiac dysrhythmias can be associated with severe blood pressure abnormalities. Coma may develop, and convulsions may be a terminal event.

Aconitum napellus (left)

Aconitum napellus, close-up of flowers (below)

Management: Fluid replacement should be instituted with respiratory support if indicated. Heart rhythm and blood pressure should be monitored and treated with appropriate medications and supportive care. Recovery is generally complete within 24 hours. Consultation with a Poison Control Center should be strongly considered. See "Poisoning by Plants with Sodium Channel Activators," p. 32.

References

Chan TYK, Tomlinson B, Critchley JAJH. Aconitine poisoning following the ingestion of Chinese herbal medicine: A report of eight cases. Aust N Z J Med 1993;23:268–271.

Lin CC, Chan TY, Deng JF. Clinical features and management of herb-induced aconitine poisoning. Ann Emerg Med 2004;43:574–579.

Martens PR, Vandevelde K. A near lethal case of combined strychnine and aconitine poisoning. J Toxicol Clin Toxicol 1993;31:133–138.

Tai YT, But PPH, Young K, et al. Cardiotoxicity after accidental herb-induced aconite poisoning. Lancet 1992;34:1254–1256.

Actaea species
Family: Ranunculaceae
Actaea pachypoda Elliott (=*Actaea alba* (L.) Mill.)
Actaea rubra (Aiton) Willd.
Actaea spicata L.

Common Names: Baneberry, Cohosh, Dolls Eyes, Herb-Christopher, Necklace Weed, Poison de Couleuvre, Pain de Couleuvre, Snakeberry, White Baneberry, White Cohosh

Description: The baneberries are perennial herbs that are 1 to 2 feet in height with large compound leaves. Small white flowers form in terminal racemes in the spring. Berry-like fruit forms in the summer or early autumn; their color depends upon the species (*Actaea pachypoda*, white; *A. rubra*, red; *A. spicata*, purplish-black).

Distribution: *Actaea* species grow throughout the north temperate zone: *Actaea pachypoda*, Nova Scotia to Georgia west to Minnesota and Missouri; *A. rubra*, Alaska to California east through Canada and the United States to the Atlantic and south through the Rockies to New Mexico; *A. spicata*, only in cultivation.

Toxic Part: Only the fruit and roots are poisonous.

Toxin: Not fully identified.

Clinical Findings: The juice exerts a direct irritant and vesicant action on the skin and mucous membranes. After ingestion, blistering and ulceration of the mucous membranes associated with salivation can occur. Nausea, vomiting, abdominal cramping, and diarrhea may occur.

Actaea pachypoda, mature fruit

Actaea spicata, fruit

Management: The irritant effect of the toxin usually limits the amount ingested. Intravenous hydration, antiemetics, and electrolyte replacement may be necessary for patients with severe gastrointestinal symptoms, particularly in children. Consultation with a Poison Control Center should be considered. See "Poisoning by Plants with Gastrointestinal Toxins," p. 28.

Actaea rubra, close-up of fruit

Actaea rubra

References

Bacon A. An experiment with the fruit of red baneberry. Rhodora 1903;5:77–79.

Hegnauer R. *Chemotaxonomie der Pflanzen, eine Übersicht über die Verbreitung und die systematische Bedeutung der Planzenstoffe.* Birkhäuser Verlag, Basel, 1962.

Adenium species
Family: Apocynaceae

Common Names: Desert Rose, Mock Azalea

Description: These shrubs or trees are 6 to 10 feet in height with swollen trunk and branches and fleshy dark green leaves with milky sap. Bright and showy funnel-form pink or purple flowers cluster at the branch tips.

Distribution: *Adenium* are cultivated as houseplants and may be grown outdoors in hot dry areas.

Toxic part: The whole plant is poisonous.

Adenium sp., flowering branch

Adenium sp., branch with flowers and leaves

Toxin: Cardioactive steroids resembling digitalis.

Clinical Findings: There are no adequately documented human poisonings, and clinical descriptions are derived primarily from animal reports. Substantial ingestion may lead to toxicity. Poisoning would be expected to produce clinical findings typical of cardioactive steroids. Toxicity has a variable latent period that depends on the quantity ingested. Dysrhythmias are usually expressed as sinus bradycardia, premature ventricular contractions, atrioventricular conduction defects, or ventricular tachydysrhythmias. Hyperkalemia, if present, may be an indicator of toxicity.

Management: Gastrointestinal decontamination as appropriate, serial electro-cardiograms, and serum potassium determinations should be performed. If serious cardioactive steroid toxicity is considered, digoxin-specific Fab should be administered. Consultation with a Poison Control Center should be considered. See "Poisoning by Plants with Cardioactive Steroids/Cardiac Glycosides," p. 24.

References

Omino EA, Kokwaro JO. Ethnobotany of Apocynaceae species in Kenya. J Ethnopharmacol 1993;40:167–180.

Yamauchi T, Abe F. Cardiac glycosides and pregnanes from *Adenium obesum*: studies on the constituents of *Adenium*. I. Chem Pharm Bull 1990;38:669–672.

Adonis species
Family: Ranunculaceae
Adonis aestivalis L.
Adonis amurensis Regel and
    Radde
Adonis annua L.
Adonis vernalis L.

Common Names: Flor de Adonis, Gota de Sangre, **Pheasant's Eye**, Red Morocco

Description: These small herbaceous perennial plants have feathery leaves, thick roots, and conspicuous yellow or crimson spring flowers that are 2 to 3 inches across on short stems.

Adonis amurensis (above)

Adonis amurensis, close-up of flower (below)

Distribution: *Adonis* species are found primarily in cultivation, usually in rock gardens, in the north temperate zones of the United States. They are alpine plants from southern Europe, Asia Minor, and Asia.

Toxic Part: The whole plant is poisonous.

Adonis annua

Toxin: Cardioactive steroids resembling digitalis.

Clinical Findings: There are no adequately documented human poisonings, and clinical descriptions are derived primarily from animal reports. Substantial ingestion may lead to toxicity. Poisoning would be expected to produce clinical findings typical of cardioactive steroids. Toxicity has a variable latent period that depends on the quantity ingested. Dysrhythmias are usually expressed as sinus bradycardia, premature ventricular contractions, atrioventricular conduction defects, or ventricular tachydysrhythmias. Hyperkalemia, if present, may be an indicator of toxicity.

Management: Gastrointestinal decontamination as appropriate, serial electrocardiograms, and serum potassium determinations should be performed. If serious cardioactive steroid toxicity is considered, digoxin-specific Fab should be administered. Consultation with a Poison Control Center should be considered. See "Poisoning by Plants with Cardioactive Steroids/Cardiac Glycosides," p. 24.

References

Cheung K, Hinds JA, Duffy P. Detection of poisoning by plant-origin cardiac glycoside with the Abbott TDx analyzer. Clin Chem 1989;35:295–297.

Davies RL. Whyte PB. *Adonis microcarpa* (pheasant's eye) toxicity in pigs fed field pea screenings. Aust Vet J 1989;66:141–143.

Aesculus species
Family: Hippocastanaceae
Aesculus californica (Spach) Nutt.
Aesculus × *carnea* Hayne
Aesculus flava Sol. (=*A. octandra* Marsh)
Aesculus glabra Willd.
Aesculus hippocastanum L.
Aesculus pavia L.
Aesculus sylvatica Bartr.

Common Names: Buckeye, Bongay, Conquerors, Fish Poison, **Horse Chestnut**, Marronnier, Marronnier d'Inde

Description: There are about 13 species of *Aesculus;* they are large trees or shrubs. These plants have palmately compound leaves, usually with five leaflets and showy clusters of flowers, which may be white, white blotched with red and yellow, pink, yellow and red, or red. The seedpod is leathery and may be smooth or warty. The common name, buckeye, is derived from the seed, which is glossy brown with a white scar.

Distribution: These plants occur primarily in the central and eastern temperate zone to the Gulf Coast and California, southern Ontario, southwestern Quebec, and Newfoundland.

Toxic Part: The seeds and twigs are poisonous. People unfamiliar with the toxicity of these species could ingest the seeds, thinking them a form of edible chestnut.

Toxin: Aescin, a complex mixture of saponins.

Clinical Findings: Most ingestions result in little or no toxicity. The saponins are poorly absorbed, but with large exposures, gastrointestinal effects including nausea, vomiting, abdominal cramping, and diarrhea may occur. Allergic sensitization to this plant is common and can cause severe allergic reactions.

Management: If severe gastrointestinal symptoms occur, intravenous hydration, antiemetics, and electrolyte replacement may be necessary for patients with severe gastrointestinal symptoms, particularly in children. Consultation with a Poison Control Center should be considered. See "Poisoning by Plants with Gastrointestinal Toxins," p. 28.

Aesculus californica, flower

Aesculus glabra

Aesculus hippocastanum

Aesculus hippocastanum, leaves and seeds

References

Jaspersen-Schib R, Theus L, Guirguis-Oeschger M, et al. Serious plant poisonings in Switzerland 1966–1994. Case analysis from the Swiss Toxicology Information Center. Schweiz Med Wochenschr 1996;126:1085–1098.

Matyunas N, Krenzelok E, Jacobsen T, et al. Horse chestnut (*Aesculus* spp.) ingestion in the United States 1985–1994. J Toxicol Clin Toxicol 1997;35:527–528.

Oschmann R, Biber A, Lang F, et al. Pharmacokinetics of beta-escin after administration of different *Aesculus* extract-containing formulations. Pharmazie 1996;51:577–581.

Popp W, Horak F, Jager S, et al. Horse chestnut (*Aesculus hippocastanum*) pollen: A Frequent cause of allergic sensitization in urban children. Allergy 1992;47:380–383.

Zhang Z, Koike K, Jia Z, Nikaido T, Guo D, Zheng J. New saponins from the seeds of *Aesculus chinensis*. Chem Pharm Bull (Tokyo) 1999;47:1515–1520.

Aethusa cynapium **L.**
Family: Umbelliferae (Apiaceae)

Common Names: Dog Parsley, Dog Poison, False Parsley, Fool's Cicely, **Fool's Parsley**, Lesser Hemlock, Small Hemlock

Description: This carrot-like plant is 8 to 24 inches high. The leaves resemble parsley but have a glossy shine on both sides and an unpleasant garlic-like odor. The white flowers and seedpods are inconspicuous and are formed on the stem tips. As the common name suggests, this plant may be consumed if mistaken for parsley.

Distribution: *Aethusa* is naturalized from Europe and grows in waste places in the extreme northern United States from Minnesota to Maine and south to Delaware, Pennsylvania, and Ohio, and in southwestern British Columbia, Ontario, Quebec, New Brunswick, and Nova Scotia.

Toxic Part: The whole plant is poisonous.

Toxin: Unsaturated aliphatic alcohols (e.g., aethusanol A) closely related to cicutoxin (from *Cicuta* species) and traces of coniine.

Aethusa cynapium

Clinical Findings: Ingestion can cause nausea, vomiting, diaphoresis, and headache. Toxicity resembles poisoning from cicutoxin (see *Cicuta* species). However, the concentration of toxin is insufficient to cause serious effects in most cases. If poisoning occurs, onset of effect is rapid, usually within 1 hour of ingestion. Symptoms include nausea, vomiting, salivation, and trismus. Generalized seizures also may occur. Death may occur if seizures do not terminate.

Management: If toxicity develops, supportive care—including airway management and protection against rhabdomyolysis and associated complications (e.g., electrolyte abnormalities and renal insufficiency)—is the mainstay of therapy. Rapidly acting anticonvulsants (i.e., diazepam or lorazepam) for persistent seizures may be needed. Consultation with a Poison Control Center should be considered. See "Poisoning by Plants with Convulsive Poisons (Seizures)," p. 25.

References

Bohlmann F, Mannhard HJ. Polyenes from *Aethusa cynapium* L. Chem Ber 1960;93:981–987.

Teuscher E, Greger H, Adrian V. Investigation of the toxicity of *Aethusa cynapium* L. (Hundspetersilie). Pharmazie 1990;45:537–538.

Aleurites species
Family: Euphorbiaceae
Aleurites cordata (Thunb.) Steud.
Aleurites fordii Hemsl.
Aleurites moluccana (L.) Willd.
Aleurites montana (Lour.) E.H. Wilson
Aleurites trisperma Blanco

Common Names:
Aleurites cordata: **Japan Oil Tree**
Aleurites fordii: Aceite, Chinawood Oil Tree, **Tung Nut**, Tung Oil Tree
Aleurites moluccana: Avellano, Candleberry, **Candlenut**, Country Walnut, Indian Walnut, Jamaican Walnut, Otaheite Walnut, Kukui, Lumbang, Nogal de la India, Noix de Bancoul, Noisette des Grands-Fonds, Noix des Molluques, Nuez de la India, Nuez Nogal, Palo de Nuez, Raguar
Aleurites montana: Mu, **Mu Oil Tree**
Aleurites trisperma: **Banu-calad**

Aleurites moluccana, fruiting branch

Description: These medium-sized trees have a milky latex, a characteristic of this plant family. The large simple leaves are often lobed. Flowers occur in large clusters and are white or white with red or orange veins. The fruit of *Aleurites fordii* is green turning to brown at maturity; the capsule is 2 to 3 inches long and contains three to seven hard seeds with a white flesh. The unhulled seed resembles an unshelled hickory nut; the hulled seed looks like a chestnut.

Distribution: *Aleurites* species grow in the West Indies, states bordering the Gulf of Mexico, Hawaii, and Guam.

Toxic Part: The whole plant is poisonous, but seeds are involved most often in human exposures.

Toxin: A derivative of phorbol, an irritant.

Clinical Findings: A sensation of discomfort, warmth, and nausea develops shortly after ingestion. This may be followed by vomiting, abdominal cramping, diarrhea, dehydration, and electrolyte imbalance.

Management: Intravenous hydration, antiemetics, and electrolyte replacement may be necessary for patients with severe gastrointestinal symptoms, particularly in children. Consultation with a Poison Control Center should be considered. See "Poisoning by Plants with Gastrointestinal Toxins," p. 28.

References

Balthrop E, Gallagher WB, McDonald TF, Camariotes S. Tung nut poisoning. J Fla Med Assoc 1954;40:813–820.

Lin TJ, Hsu CI, Lee KH, et al. Two outbreaks of acute Tung Nut (*Aleurites fordii*) poisoning. J Toxicol Clin Toxicol 1996;34:87–92.

Allamanda cathartica L.
Family: Apocynaceae

Common Names: **Allamanda**, Canario, Cautiva, Flor de Barbero, Lani-Ali'i, Nani-Ali'i, Yellow Allamanda

Description: This ornamental sprawling shrub or woody climber has large yellow flowers and opposite lance-shaped leaves that are 4 to 6 inches long. *Allamanda cathartica* usually is propagated by cuttings, which do not fruit. *Allamanda neriifolia* Hook., cultivated in the same geographic area, is a bush with more diminutive leaves and flowers; it frequently fruits. The stems and leaves contain irritating latex.

Distribution: *Allamanda* species are commonly cultivated in the West Indies, Florida, and Hawaii, and elsewhere in the tropics. In the temperate zone, it is sometimes found as a houseplant. *Allamanda cathartica* was introduced from Brazil and, as the name implies, was once used as a cathartic.

Toxic Part: Bark, leaves, fruit, seeds, and sap are poisonous.

Toxin: Plumericin, a weak gastrointestinal irritant.

Allamanda cathartica, close-up of flowering branch

Allamanda cathartica

Clinical Findings: Exposure may cause abdominal cramping and diarrhea. Large exposures may cause dehydration and electrolyte abnormalities.

Management: Most cases require no therapy. Intravenous hydration, antiemetics, and electrolyte replacement may be necessary for patients with severe gastrointestinal symptoms, particularly in children. Consultation with a Poison Control Center should be considered. See "Poisoning by Plants with Gastrointestinal Toxins," p. 28.

References

Abdel-Kader MS, Wisse J, Evans R, et al. Bioactive iridoids and a new lignan from *Allamanda cathartica* and *Himatanthus fallax* from the Suriname rainforest. J Nat Prod 1997;60:1294–1297.

Allium species
Family: Liliaceae

Common Names:

Allium canadense L.: **Wild Garlic**, Ail du Canada, Meadow Garlic, Meadow Rose Leek, Onion Tree, **Wild Onion**.

Allium cepa L.: 'Aka'akai, **Onion**, Zongnon

Allium sativum L.: Ajo, 'Aka'akai-pilau, Clown Treacle, **Garlic**, Lai, Poor Man's Treacle

Allium vineale L.: **Field Garlic**, Crow Garlic

Allium canadense

Allium canadense, young bulbs

Description: There are about 400 species of *Allium*, each with a strong-smelling bulb when it is crushed or scraped. The plants are perennial, with long, slender leaves that are flat or tubular. Flowers are white to pink or purple. The species listed above are most often implicated in human toxicity.

Distribution: There are many cultivated and native species of *Allium* throughout the United States, including Hawaii and Alaska. *Allium canadense* may be found from Minnesota to Ontario, south to Texas, east to Quebec and western New Brunswick, and south to Florida.

Toxic Part: Bulbs, bulblets, flowers, and stems are "poisonous" when consumed in large quantities, but onion (*A. cepa*) and garlic (*A. sativum*) are of course grown commercially and consumed in smaller quantities for the most part with no ill effects.

Toxin: *N*-Propyl sulfide, methyl disulfide, and allyl disulfide.

Clinical Findings: Most exposures lead to no or minimal symptoms. Substantial ingestion can cause nausea, vomiting, abdominal cramping, diarrhea, and dehydration. Platelet aggregation is inhibited, and the current practice of ingesting large amounts of garlic supplements for cancer prevention and cardiovascular benefits may lead to bleeding. Chronic ingestion of the bulblets may diminish iodine uptake by the thyroid.

Management: Most cases require no therapy. Intravenous hydration, antiemetics, and electrolyte replacement may be necessary for patients with severe gastrointestinal symptoms, particularly in children. Bleeding, depending on its location, may require specific diagnostic testing and therapy. Consultation with a Poison Control Center should be considered. See "Poisoning by Plants with Gastrointestinal Toxins," p. 28.

References

Amagase H, Petesch BL, Matsuura H, Kasuga S, Itakura Y. Intake of garlic and its bioactive components. J Nutr 2001;131:955S–962S.

Ariga T, Tsuj K, Seki T, Moritomo T, Yamamoto JI. Antithrombotic and antineoplastic effects of phyto-organosulfur compounds. Biofactors 2000;13:251–255.

Le Bon AM, Siess MH. Organosulfur compounds from *Allium* and the chemoprevention of cancer. Drug Metabol Drug Interact 2000;17:51–79.

Alocasia species
Family: Araceae

Common Names: Ahe Poi, 'Ape, Cabeza de Burro, Chine Ape, **Elephant's Ear**, Malanga Cara de Chivo, Malanga de Jardín, Papao-Apaka, Papao-Atolong, Taro

Description: These erect perennials have single, long-stemmed, spearhead-shaped leaves that are prominently veined and often varicolored. Flowers appear on a spadix subtended by a greenish spathe similar to *Colocasia.* Individual plants may develop from runners (rhizomes).

Distribution: Species of the genus *Alocasia* are popularly cultivated throughout the United States and are hardy in tropical and subtropical areas. Tubers of some species of *Alocasia* are eaten after cooking.

Alocasia cv. 'Hilo's Beauty'

Alocasia macrorrhiza cv. 'Variegata'

Alocasia × amazonica

Alocasia watsoniana

Toxic Part: The leaves, stems, and tubers may be injurious.

Toxin: Raphides of water-insoluble calcium oxalate and unverified proteinaceous toxins.

Clinical Findings: A painful burning sensation of the lips and mouth result from ingestion. There is an inflammatory reaction, often with edema and blistering. Hoarseness, dysphonia, and dysphagia may result.

Management: The pain and edema recede slowly without therapy. Cool liquids or demulcents held in the mouth may bring some relief. Analgesics may be indicated. The insoluble oxalate in these plants does not cause systemic oxalate poisoning. Consultation with a Poison Control Center should be considered. See "Poisoning by Plants with Calcium Oxalate Crystals," p. 23.

References

Chan TY, Chan LY, Tam LS, Critchley JA. Neurotoxicity following the ingestion of a Chinese medicinal plant, *Alocasia macrorrhiza*. Hum Exp Toxicol 1995;14:727–728.

Franceschi VR, Nakata PA. Calcium oxalate in plants: Formation and function. Annu Rev Plant Biol 2005;56:41–71.

Lin TJ, Hung DZ, Hu WH, Yang DY, Wu TC, Deng JF. Calcium oxalate is the main toxic component in clinical presentations of *Alocasia macrorrhiza* (L.) Schott and Endl. poisonings. Vet Hum Toxicol 1998;40:93–95.

Rauber A. Observations on the idioblasts of *Dieffenbachia*. J Toxicol Clin Toxicol 1985; 23:79–90.

Aloe species
Family: Liliaceae

Common Names: **Aloe**, L'Aloé, Panini-'Awa' Awa, Star Cactus, Sábila, Sempervivum, Sinkam Bible, Zábila, Závila

Description: There are about 250 species of *Aloe*, which are perennial herbs. The leaves usually appear in compact rosettes and are fleshy, thick, hard, and pointed; they may have teeth or spines on the margins. The flowers are usually red or yellow. Seedpods split on drying.

Distribution: Aloes are cultivated extensively in the West Indies, Florida, Texas, southern California, Hawaii, and Guam. They commonly appear as a summer planting or houseplant. Some aloes, for example, *Aloe vera* (L.) Burm. f. (=*A. barbadensis* Mill.), known as Barbados Aloe, Curaçao Aloe, Medicinal Aloe, or Unguentine Cactus, are grown commercially for the cathartic glycosides in the latex and for their mucilaginous gel, which is employed as a skin moistener. Although *Aloe* is sold in health food stores as an intestinal tonic, excessive oral consumption can cause symptoms of toxicity.

Toxic Part: The latex juice within the leaves.

Toxin: Barbaloin, an anthraquinone glycoside.

Aloe speciosa, flower (right)

Aloe vera (below)

Clinical Findings: Most exposures cause minimal or no toxicity. Large exposures may cause gastrointestinal irritation. Some references suggest that aloe can cause nephritis, or kidney inflammation, although no human cases support this purported effect. Contact dermatitis and allergic reactions can occur with exposure to this plant.

Management: Most cases require no therapy. In severe cases, intravenous hydration, antiemetics, and electrolyte replacement may be necessary, particularly in children. Consultation with a Poison Control Center should be considered. See "Poisoning by Plants with Gastrointestinal Toxins," p. 28.

References

Hunter D, Frumkin A. Adverse reactions to vitamin E and *Aloe vera* preparations after dermabrasion and chemical peel. Cutis 1991;47:193–196.

Ishii Y, Tanizawa H, Takino Y. Studies of aloe. III. Mechanism of cathartic effect. (2). Chem Pharm Bull (Tokyo) 1990;38:197–200.

Ishii Y, Takino Y, Toyo'oka T, Tanizawa H. Studies of aloe. VI. Cathartic effect of isobarbaloin. Biol Pharm Bull 1998;21:1226–1227.

Spoerke DG, Elkins BR. *Aloe vera*—fact or quackery? Vet Hum Toxicol 1980;22:418–424.

Amaryllis species
Hippeastrum species
Family: Amaryllidaceae

Common Names: Amaryllis (*Amaryllis* species of horticulture, except *A. belladonna* L., are species of *Hippeastrum*), Azucena, Barbados Lily, Belladonna Lily, Cape Belladonna, Lirio, Naked Lady Lily, Saint Joseph Lily, Tararaco

Hippeastrum cv. 'Basuto,' close-up of flowers (left)

Hippeastrum sp., close-up of flowers (below)

Hippeastrum puniceum

Description: *Amaryllis* and *Hippeastrum* have variously colored flowers, but shades of red or pink with white are most common. Bulbs usually extend partly above ground. The leaves are shiny, thick, and strap shaped.

Distribution: These plants occur as cultivated ornamentals only. They are extremely popular as a summer garden plant in temperate zones and as a potted plant throughout the year.

Toxic Part: Bulbs are poisonous.

Toxin: Lycorine and related phenanthridine alkaloids (see *Narcissus*).

Clinical Findings: Ingestion of small amounts produces few or no symptoms. Large exposures may cause nausea, vomiting, abdominal cramping, diarrhea, dehydration, and electrolyte imbalance.

Management: Most exposures result in minimal or no toxicity. Intravenous hydration, antiemetics, and electrolyte replacement may be necessary for patients with severe gastrointestinal symptoms, particularly in children. Consultation with a Poison Control Center should be considered. See "Poisoning by Plants with Gastrointestinal Toxins," p. 28.

References

Mugge C, Schablinski B, Obst K, Dopke W. Alkaloids from *Hippeastrum* hybrids. Pharmazie 1994;49;444–447.

Anemone species
Pulsatilla species
Family: Ranunculaceae
Anemone canadensis L.
Anemone coronaria L.
Pulsatilla patens (L.) Mill. (=*A. nuttalliana* DC.; *A. patens* L.)
Pulsatilla vulgaris Mill. (=*A. pulsatilla* L.)

Common Names: Anemone, April Fools, Cat's Eyes, Gosling, Hartshorn Plant, Lily of the Field, Lion's Beard, Nightcaps, Nimble Weed, **Pasque Flower** (*Pulsatilla* species), Prairie Crocus, Prairie Hen Flower, Prairie Smoke, Thimbleweed, Wild Crocus, Windflower

Anemone coronaria

Anemone canadensis

Pulsatilla patens, close-up of flowers

Description: These perennial herbs are often used in rock gardens because they are usually no more than 1 foot high. The divided flowers appear singly on the stem and may be showy yellow, red, white, or purple. *Anemone* is also sold as a cut flower.

Distribution: There are many species of *Anemone* in Canada, Alaska, and the north temperate zone of the United States. *Pulsatilla vulgaris* and *P. patens* are of European origin.

Toxic Part: The whole plant is poisonous.

Pulsatilla vulgaris cv. 'Papageno' (above)

Pulsatilla vulgaris, close-up of flower (below)

Toxin: Protoanemonin, an irritant.

Clinical Findings: There are no adequately documented human poisonings, and clinical descriptions are derived primarily from animal reports. Intense pain and inflammation of the mouth with blistering, ulceration, and profuse salivation can occur. Bloody emesis and diarrhea develop in association with severe abdominal cramps. Central nervous system involvement is manifested by dizziness, syncope, and seizures.

Management: Most exposures result in minimal or no toxicity. Intravenous hydration, antiemetics, and electrolyte replacement may be necessary for patients with severe gastrointestinal symptoms, particularly in children. Consultation with a Poison Control Center should be considered. See "Poisoning by Plants with Gastrointestinal Toxins," p. 28.

If seizures occur, rapidly acting anticonvulsants, such as intravenous diazepam, should be utilized along with other supportive measures. Consultation with a Poison Control Center should be strongly considered. See "Poisoning by Plants with Convulsant Posisons (Seizures) p. 25.

References

Aaron TH, Muttitt EL. Vesicant dermatitis due to prairie crocus (*Anemone patens* L.). Arch Dermatol 1964;90:168–171.

Turner NJ. Counter-irritant and other medicinal uses of plants in Ranunculaceae by native peoples in British Columbia and neighbouring areas. J Ethnopharmacol 1984;11: 181–201.

Vance JC. Toxic plants of Minnesota: skin toxicity of the Prairie Crocus (*Anemone patens* L.). Minn Med 1982;65:149–151.

Anthurium species
Family: Araceae

Common Names: Anthurium, Anturio, Crystal Anthurium, Flamingo Flower, Flamingo Lily, Flor de Culebra, Hoja Grande, Lengua de Vaca, Lombricero, Oilcloth Flower, Pigtail Plant, Strap Flower, Tail Flower

Description: The most common species of *Anthurium* grow about 2 feet high and have dark green, heart-shaped, leathery leaves. The spadix is a persistent scarlet, white, or green spike, subtended by a showy spathe that is white, pink, or red in color. Flowering may be followed by colorful, showy berries. Other species, particularly wild forms, have shiny, broad leaves that grow to 3 feet in length. In these types, the floral bracts (spathes) are usually green or maroon and less conspicuous.

Anthurium × ferrierense, flower

Distribution: These plants are native to tropical America. They are houseplants in most areas, but may be grown in gardens in south Florida and Hawaii.

Toxic Part: Leaves and stems are injurious.

Toxin: Raphides of water-insoluble calcium oxalate and unverified proteinaceous toxins.

Clinical Findings: A painful burning sensation of the lips and oral cavity results from ingestion. There is an inflammatory reaction,

Anthurium × *roseum* (above)

Anthurium wildenowii (right)

often with edema and blistering. Hoarseness, dysphonia, and dysphagia may result.

Management: The pain and edema recede slowly without therapy. Cool liquids or demulcents held in the mouth may bring some relief. Analgesics may be indicated. The insoluble oxalate in these plants does not cause systemic oxalate poisoning. Consultation with a Poison Control Center should be considered. See "Poisoning by Plants with Calcium Oxalate Crystals," p. 23.

Reference

Rauber A. Observations on the idioblasts of *Dieffenbachia*. J Toxicol Clin Toxicol 1985;23:79–90.

Arisaema species
Family: Araceae
Arisaema dracontium (L.) Schott
Arisaema triphyllum (L.) Schott (includes subsp. *stewardsonii* (Britton) Huttl. and subsp. *triphyllum* [=*A. atrorubens* (Aiton) Blume])

Common Names:
Arisaema dracontium: Dragon Arum, Dragon Root, Dragon Tail, Dragon Head, **Green Dragon**, Jack-in-the-Pulpit
Arisaema triphyllum: Bog Onion, Brown Dragon, Cuckoo Plant, Devil's Ear, Indian Jack-in-the-Pulpit, Indian Turnip, **Jack-in-the-Pulpit**, Memory Root, Pepper Turnip, Petit Précheur, Priests Pentle, Small Jack -in-the-Pulpit, Starchwort, Three Leaved Indian Turnip, Wake Robin

Description: *Arisaema dracontium* has a single 3-foot stalk with 7 to 13 leaflets at the tip. Several inches below the leaflets, a single green "flower" (spathe) up to 3 inches long emerges from the stalk. This spathe has a slender, funnel-shaped, protruding hood that extends backward. The spikelike, erect spadix inside extends for several inches. Berries are orange-red when ripe. *Arisaema triphyllum* has two leaves branching from the stalk with three lance-shaped leaflets at the end. The spathe appears between the branching of the two leaves and may be green, green striped with brown, or brown. It is shaped like a pulpit with a long pointed hood extending over the short blunt spadix. Berries

Arisaema dracontium

are red when ripe. *Arisaema* species bloom from late spring to early autumn, and grow from swollen underground stems (bulbs).

Distribution: *Arisaema* species grow in damp shady areas from Quebec to Florida west to Ontario and southwest to Minnesota and Texas.

Toxic Part: Whole plant.

Toxin: Raphides of water-insoluble calcium oxalate and unverified proteinaceous toxins.

Clinical Findings: A painful burning sensation of the lips and mouth results from ingestion. There is an inflammatory reaction, often with edema and blistering. Hoarseness, dysphonia, and dysphagia may result.

Management: The pain and edema recede slowly without therapy. Cool liquids or demulcents held in the mouth may bring some relief. Analgesics may be indicated. The insoluble oxalate in these plants does not cause systemic oxalate poisoning. Consultation with a Poison Control Center should be considered. See "Poisoning by Plants with Calcium Oxalate Crystals," p. 23.

Arisaema triphyllum, close-up of flower

References

Rauber A. Observations on the idioblasts of *Dieffenbachia*. J Toxicol Clin Toxicol 1985;23:79–90.

Arum species
Family: Araceae
Arum italicum Mill.
Arum maculatum L.
Arum palaestinum Boiss.

Common Names:
Arum italicum: **Italian Arum**, Italian Lords-and-Ladies
Arum maculatum: Adam and Eve, Bobbins, **Cuckoopint**, Flekkmunkehette, Lords-and-Ladies, Munkehette, Starch Root
Arum palaestinum: Arum Sanctum, **Black Calla**, Solomon's Lily

Description: Arums are stemless plants with 8- to 10-inch ovate leaves and tuberous roots. The showy "flower," a spathe, is generally dullish purple in the above species and encloses a spike (spadix) on which brilliant red fruits form.

Distribution: These common houseplants are hardy outdoor plants in the South. All these species originated in Europe and the Near East.

Toxic Part: The whole plant is injurious. Most exposures occur because of ingestion of the colorful fruit that attracts children.

Toxin: Poorly identified gastrointestinal toxin. Previously thought to contain calcium oxalates; this has not been clearly established.

Clinical Findings: Mucous membrane irritation, burning sensation, and possible ulceration. Nausea, vomiting, abdominal cramping, diarrhea, and dehydration can occur with large exposures.

Arum italicum, leaf

Arum italicum, unripe fruit

Arum italicum, ripe fruit

Management: Oral lesions are treated with cool liquids and possibly analgesics. Intravenous hydration, antiemetics, and electrolyte replacement may be necessary for patients with severe gastrointestinal symptoms, particularly in children. Consultation with a Poison Control Center should be considered. See "Poisoning by Plants with Gastrointestinal Toxins," p. 28.

References

Jaspersen-Schib R, Theus L, Guirguis-Oeschger M, Gossweiler B, Meier-Abt PJ. Serious plant poisonings in Switzerland 1966–1994. Case analysis from the Swiss Toxicology Information Center. Schweiz Med Wochenschr 1996;126:1085–1098.

Stahl E, Kaitenbach U. Die basischen Inhaltsstoffe des Aronstabes (*Arum maculatum* L.) Arch Pharm (Weinheim) 1965;298:599.

Astralagus species
Family: Leguminosae (Fabaceae)

Astralagus layneae Greene
Astralagus calycosus Torr. ex. S. Wats.
Astralagus lentiginosus Dougl. ex. Hook.
Astralagus mollissimus Torr.
Astralagus wootonii Sheldon

Astralagus mollissimus, flowers and fruit

Common Names:
Astralagus layneae: **Layne's Locoweed**
Astralagus calycosus: **King's Locoweed**
Astralagus lentiginosus: **Speckled Loco**, Freckled Milkvetch
Astralagus mollissimus: **WoolyLoco**, Purple Loco
Astralagus wootonii: **Wooton Loco**, Garbancillo, Wooton Milkvetch

Description: *Astragalus* comprises about 1,000 species of herbaceous plants. The leaves are alternate, pinnate, with entire leaflets; some species are covered with silvery hairs. Depending on the species, flowers can be white, pink, red, purple, or yellow, with five petals, three upper ones and two lower fused ones. Fruits are fleshy or papery legumes.

Distribution: This genus is found in many areas of the Northern Hemisphere, especially in the arid western United States. Some species of European, African, or Asian origin have been introduced into cultivation.

Toxic Part: Whole plant.

Toxin: Swainsonine, an indolizidine alkaloid that inhibits lysosomal alpha-mannosidase.

Astralagus wootonii, leaves and fruit

Clinical Findings: There are no adequately documented human poisonings, and clinical descriptions are derived primarily from animal reports. *Astragalus membranaceus* is used in herbal medicine for the treatment of viral myocarditis. In animals, swainsonine produces a lysosomal storage disease that manifests as locoism. Locoism includes hyperactive and aggressive behavior, a stiff and clumsy gait, seizures, increasing miscoordination, weakness, and death. Locoism may also develop in livestock that feed on *Oxytropis* sp., *Sida carpinifolia,* and *Ipomoea carnea,* plants that also contain swainsonine.

Management: Symptomatic and supportive care. Consultation with a Poison Control Center should be considered.

References

James LF, Panter KE, Gaffield W, Molyneux RJ. Biomedical applications of poisonous plant research. J Agric Food Chem 2004;52:3211–3230.

Pfister JA, Stegelmeier BL, Gardner DR, James LF. Grazing of spotted locoweed (*Astragalus lentiginosus*) by cattle and horses in Arizona. J Anim Sci 2003;81:2285–2293.

Atropa belladonna L.
Family: Solanaceae

Common Names: Belladonna, Black Nightshade, **Deadly Nightshade,** Nightshade, Sleeping Nightshade

Description: These perennial plants are about 3 feet high and are often cultivated in flower gardens. The stems are very branched with 6-inch ovate leaves. Solitary flowers, which emerge from the leaf axils, are blue-purple to dull red and about 1 inch long. The fruit is nearly globular, about 0.5 inch in diameter, and is purple to shiny black when mature. The root is a thick rhizome. The sap is reddish.

Toxic Part: The whole plant is toxic.

Atropa belladonna (above)

Atropa belladonna, mature fruit (below)

Toxin: Atropine, scopolamine, and other anticholinergic alkaloids.

Clinical Findings: Intoxication results in dry mouth with dysphagia and dysphonia, tachycardia, and urinary retention. Elevation of body temperature may be accompanied by flushed, dry skin. Mydriasis, blurred vision, excitement and delirium, headache, and confusion may be observed.

Management: Initially, symptomatic and supportive care should be given. If the severity of the intoxication warrants intervention (hyperthermia, delirium), an antidote, physostigmine, is available. Consultation with a Poison Control Center should be considered. See "Poisoning by Plants with Anticholinergic (Antimuscarinic) Poisons," p. 21.

References

Caksen H, Odabas D, Akbayram S, et al. Deadly nightshade (*Atropa belladonna*) intoxication: An analysis of 49 children. Hum Exp Toxicol 2003;22:665–668.

Joshi P, Wicks AC, Munshi SK. Recurrent autumnal psychosis. Postgrad Med J 2003;79: 239–240.

Schneider F, Lutun P, Kintz P, et al. Plasma and urine concentrations of atropine after the ingestion of cooked deadly nightshade berries. J Toxicol Clin Toxicol 1996;34:113–117.

Trabattoni G, Visintini D, Terzano GM, et al. Accidental poisoning with deadly nightshade berries: A case report. Hum Toxicol 1984;3:513–516.

Aucuba japonica Thunb.
Family: Cornaceae

Common Names: Japanese Aucuba, Japanese Laurel, Spotted Laurel

Description: This evergreen usually is grown as a large bush. The opposite leaves are about 7 inches long and coarsely toothed. There are numerous horticultural varieties that differ in leaf shape and color. Purple flowers are borne in panicles at the ends of branches. The fruit is a scarlet berry, matures in early winter, and contains a single seed.

Aucuba japonica

Distribution: This plant is cultivated as an ornamental in the Atlantic Gulf states from Texas to Washington, DC, and along the Pacific Coast, and is also a houseplant.

Toxic Part: Although the toxin is present in every part of the plant, the majority of exposures involve the colorful fruit.

Toxin: Aucubin, an acid-labile glycoside.

Clinical Findings: Most exposures cause minimal or no toxicity. Large exposures can cause nausea, vomiting, and abdominal cramping.

Aucuba japonica, immature fruit
Aucuba japonica

Management: Most cases require no therapy. Intravenous hydration, anti-emetics, and electrolyte replacement may be necessary for patients with severe gastrointestinal symptoms, particularly in children. Consultation with a Poison Control Center should be considered. See "Poisoning by Plants with Gastrointestinal Toxins," p. 28.

References

Chang LM, Yun HS, Kim YS, Ahn JW. Aucubin: potential antidote for alpha-amanitin poisoning. J Toxicol Clin Toxicol 1984;22:77–85.

Leaveau AM, Durand M, Paris RR. Sur la toxicité des fruits de l'*Aucuba japonica* (Cornacees). Plant Med Phytother 1979;13:199.

Baptisia species
Family: Leguminosae (Fabaceae)

Common Names: Cloverbloom, False Indigo, Horse Flea Weed, Horsefly, Horsefly Weed, Indigo Weed, Prairie Indigo, Rattlebush, Rattleweed, Shoofly, **Wild Indigo**, Yellow Indigo

Baptisia cv. 'Purple Smoke,' flowers

Baptisia tinctoria

Description: There are about 30 species of *Baptisia*. These herbs grow to about 3 feet and have oval leaflets. The sweetpea-like flowers are blue, yellow, or white and grow in racemes. The fruit is a pealike pod.

Distribution: *Baptisia* species are native throughout the United States from Texas, Missouri, and Minnesota east to the Atlantic coast, south to Florida, and north to Massachusetts. In Canada, the genus occurs only in southern Ontario.

Toxic Part: The entire plant is toxic.

Toxin: Cytisine and other related nicotine-like alkaloids.

Clinical Findings: Initial gastrointestinal effects may be followed by those typical of nicotine poisoning; these include hypertension, large pupils, sweating, and perhaps seizures.

Baptisia alba, flower

Severe poisoning produces coma, weakness, and paralysis that may result in death due to respiratory failure.

Management: Symptomatic and supportive care should be given, with attention to adequacy of ventilation and vital signs. Atropine may reverse some of the toxic effects. Consultation with a Poison Control Center should be strongly considered. See "Poisoning by Plants with Nicotine-like Alkaloids," p. 30.

References

Daly JW. Nicotinic agonists, antagonists, and modulators from natural sources. Cell Mol Neurobiol 2005;25:513–552.

Klocking H-P, Richter M, Damm G. Pharmacokinetic studies with ³H-cytisine. Arch Toxicol 1980;4(suppl):312–314.

Wagner H, Horhammer L, Budweg W, et al. Investigation of the glycosides from *Baptisia tinctoria*. 3. Synthesis of pseudobaptisins. Chem Ber 1969;102:3006–3008.

Blighia sapida K.D. Koenig
Family: Sapindaceae

Common Names: Ackee, Akee, Aki, Arbre Fricassé, Isin, Ishin, Seso Vegetal

Description: The ackee is a tree growing about 30 to 40 feet in height. It has compound leaves with five pairs of leaflets; the longest is about 6 inches at the tip. There are small, greenish-white flowers. A conspicuous reddish fruit splits at maturity to reveal three shiny black seeds embedded in a snowy white waxy aril.

Distribution: *Blighia* grows in the West Indies, Florida, and Hawaii.

Toxic Part: In the unripe fruit, the toxin is concentrated in the seeds and, especially, in the unripe aril that surrounds the seeds. Consumption of the aril from unripe fruit has resulted in fatalities. When ripe, Akee is a traditional food in Jamaica, where the aril is cooked with codfish.

Blighia sapida, fruit

Toxin: Hypoglycin A, a cyclopropyl amino acid. This toxin and its metabolites inhibit fatty acid metabolism and gluconeogenesis, leading to elevated dicarboxylic acid levels, hypoketonemia, and liver failure (microvesicular steatosis similar to Reye's syndrome).

Clinical Findings: Ingestion of the unripe fruit gives rise to "Jamaican vomiting sickness," a condition of severe vomiting that usually occurs shortly after ingestion. Lethargy, paraesthesias, hypoglycemia, hypokalemia, metabolic acidosis, seizures, and coma may occur. With chronic exposure, hepatotoxicity may develop as a result of the altered fatty acid metabolism. Mortality may occur from either acute or chronic poisoning. Confirmation of suspected cases can be made by measuring elevated dicarboxylic acid levels in urine by gas chromatography.

Blighia sapida, open fruit revealing white arils and seeds

Management: Treatment is supportive because no specific antidote exists. Patients with suspected ingestion should have early gastrointestinal decontamination, if appropriate, due to the potency of the toxin. Fluid hydration, antiemetics, electrolyte correction, glucose infusion, and supportive care should be utilized as needed. Consultation with a Poison Control Center should be considered.

References

Barennes H, Valea I, Boudat AM, Idle JR, Nagot N. Early glucose and methylene blue are effective against unripe ackee apple (*Blighia sapida*) poisoning in mice. Food Chem Toxicol 2004;42:809–815.

Chase GW Jr, Landen WO Jr, Soliman AG. Hypoglycin A content in the aril, seeds, and husks of ackee fruit at various stages of ripeness. J Assoc Off Anal Chem 1990;73: 318–319.

McTague JA, Forney R. Jamaican vomiting sickness in Toledo, Ohio. Ann Emerg Med 1994; 23:1116–1118.

Meda HA, Diallo B, Buchet JP, et al. Epidemic of fatal encephalopathy in preschool children in Burkina Faso and consumption of unripe ackee (*Blighia sapida*) fruit. Lancet 1999;353: 536–540.

Moya J. Ackee (*Blighia sapida*) poisoning in the Northern Province, Haiti, 2001. Epidemiol Bull 2001;22:8–9.

Sherratt HAS. Hypoglycin, the famous toxin of the unripe Jamaican ackee fruit. Trends Pharmacol Sci 1986;7:186–191.

Tanaka K, Kean EA, Johnson G. Jamaican vomiting sickness. N Engl J Med 1976;295: 461–467.

Brassaia actinophylla Endl.
(=*Schlefflera actinophylla* (Endl.) Harms)
Family: Araliaceae

Common Names: Australian Ivy Palm, Australian Umbrella Tree, Octopus Tree, Queensland Umbrella Tree, Queen's Umbrella Tree, **Schefflera**, Starleaf

Description: An evergreen tree growing to 40 feet, on stout stems with a few short branches at the top. Leaves are spreading, 2 to 4 feet long and glossy, usually with 7 to 16 leaflets. Flowers are small, red, and clustered, and are formed in racemes above the foliage.

Distribution: Native to Australia, New Guinea, and Java, the plant was introduced to many other tropical and subtropical regions where it is cultivated as an ornamental and naturalized into the landscape. It is also grown as a pot plant in homes and in tubs for interior landscapes such as malls.

Toxic Part: The leaves primarily. In addition, exposure to the sap from the plant or leaves may produce contact dermatitis.

Toxin: Raphides of water-insoluble calcium oxalate and unverified proteinaceous toxins.

Clinical Findings: A painful burning sensation of the lips and mouth results from ingestion. There is an inflammatory reaction, often with edema and blistering. Hoarseness, dysphonia, and dysphagia may result.

Brassaia actinophylla, branch with leaves (above)

Brassaia actinophylla, flowers (below)

Management: The pain and edema recede slowly without therapy. Cool liquids or demulcents held in the mouth may bring some relief. Analgesics may be indicated. The insoluble oxalate in these plants does not cause systemic oxalate poisoning. Consultation with a Poison Control Center should be considered. See "Poisoning by Plants with Calcium Oxalate Crystals," p. 23.

References

Franceschi VR, Nakata PA. Calcium oxalate in plants: formation and function. Annu Rev Plant Biol 2005;56;41–71.

Mitchell JC. Allergic contact dermatitis from *Hedera helix* and *Brassaia actinophylla* (Araliaceae). Contact Dermatitis 1981;7:158–159.

Rauber A. Observations on the idioblasts of *Dieffenbachia*. J Toxicol Clin Toxicol 1985;23: 79–90.

Caesalpinia species
Family: Leguminosae (Fabaceae)
Caesalpinia bonduc (L.) Roxb. (=*C. crista* L.; *C. bonducella* Flem.)
Caesalpinia drummondii (Torr. & A. Gray) Fisher
Caesalpinia gilliesii (Hook.) Dietrich (=*Poinciana gilliesii* Hook.)
Caesalpinia mexicana A. Gray
Caesalpinia pulcherrima (L.) Sw. (=*Poinciana pulcherrima* L.)
Caesalpinia vesicaria L.

Caesalpinia bonduc, close-up of flower (left)

Caesalpinia bonduc, seeds in open fruit (below)

Caesalpinia gilliesii, close-up of flower (above)

Caesalpinia pulcherrima (below)

Common Names:

Caesalpinia bonduc: Brier, **Grey Nicker Bean**, Guacalote Amarillo, Haba de San Antonio, Horse Nicker, Mato Azul, Mato de Playa, Nickal

Caesalpinia drummondii: There are no common names in our geographic area.

Caesalpinia gilliesii: Barbados Pride, **Bird of Paradise**, Brazilwood, Dwarf Ponciana, Espiga de Amor, Flower Fence, Ponciana

Caesalpinia mexicana: **Mexican Bird of Paradise**, Medican Poinciana

Caesalpinia pulcherrima: **Barbados Flower**, Barbados Pride, Dwarf Poinciana, Carzazo, Clavellina, Doddle-Do, Dul-Dul, Fench, Flor de Camarón, Flower Fence, Francillade, Guacamaya, Maravilla, 'Ohai-Ali'i, Peacock Flower, Spanish Carantion, Tabachin

Caesalpinia vesicaria: Brasil, Brasiletto, Indian Savin Tree, **Poinciana**

Description: There are about 150 species of *Caesalpinia* distributed throughout the world.

Caesalpinia bonduc: This scrambling shrub rarely grows erect. The stems have numerous prickles, and the leaves are doubly compound. Yellow flowers grow in racemes. Fruits are dark brown, 3 inches long and 1.5 inches wide, covered with prickles, and contain two flattened, globular seeds about 1 inch in diameter.

Caesalpinia drummondii: This shrublet grows about 8 inches high. Fruits are slightly less than 1 inch long and contain one or two seeds.

Caesalpinia gilliesii: This shrub is 15 feet tall and has no spines. The leaves are compound. The yellow flowers have bright red, protruding filaments.

The fruit is about 4 inches long, 0.75 inches wide, and contains six to eight seeds.

Caesalpinia mexicana: This small, spineless tree has yellow flowers. The fruit is a flat pod about 2.5 inches long.

Caesalpinia pulcherrima: This upright shrub has scattered thorns and yellow flowers with red streaks that grow in terminal racemes. The fruit is a flat pod about 4 inches long containing five to eight shiny brown, flat, beanlike seeds about 3/8 inch long.

Caesalpinia vesicaria: This small tree has an aromatic odor. Its trunk and branches are prickly. Yellow flowers are streaked with red and form in terminal racemes. The fruit is a flat pod about 4 inches long and less than 1 inch wide.

Distribution:

Caesalpinia bonduc: This weedlike plant is common in the West Indies from the Bahamas to Barbados.

Caesalpinia drummondii: This rare species is a native of southern Texas and Mexico.

Caesalpinia gilliesii: Bird of Paradise plants are cultivated extensively in Florida and the Gulf coast states to Texas and in Arizona.

Caesalpinia mexicana: This ornamental is cultivated in southern Texas and Mexico.

Caesalpinia pulcherrima: Barbados Pride is cultivated extensively in the West Indies from the Bahamas to Barbados and in Hawaii, Guam, and frost-free areas of Florida, Texas, and Arizona.

Caesalpinia vesicaria: This tree is native to the Bahamas, Cuba, Jamaica, and Haiti. It is not found in Puerto Rico.

Toxic Part: The seeds are poisonous. The immature seeds of some species (e.g., *Caesalpinia pulcherrima*) are edible; roasting the seeds of some other species (e.g., *C. bonduc*) abolishes toxicity.

Toxin: Tannins, protein precipitants that are gastrointestinal irritants.

Clinical Findings: After a latent period of 30 minutes to 6 hours, nausea, vomiting, abdominal cramping, diarrhea, and dehydration can occur.

Management: Intravenous hydration, antiemetics, and electrolyte replacement may be necessary for patients with severe gastrointestinal symptoms, particularly in children. Consultation with a Poison Control Center should be considered. See "Poisoning by Plants with Gastrointestinal Toxins," p. 28.

References

Adam F, Adam JC, Hasselot N. Intoxication par ingestion de ditakh. Med Trop 1991;51: 455–458.

Anonymous. Toxicity studies of Arizona ornamental plants. Ariz Med 1958;15:512–514.

Che CT, McPherson DD, Cordell GA, et al. Pulcherralpin, a new diterpene ester from *Caesalpinia pulcherrima*. J Nat Prod 1986;49:561–569.

Caladium bicolor (Aiton) Vent.
Family: Araceae

Common Names: Angel Wings, Caládio, **Caladium**, Cananga, Capotillo, Courer Saignant, Corazón de Cabrito, Elephant's Ear, Heart-of-Jesus, Lágrimas de María, Mother-in-Law Plant, Paleta de Pintor

Description: Caladiums have showy, heavy, variegated, prominently veined, heart-shaped leaves. Leaf coloration may vary from white to orange or red, depending on species.

Distribution: These plants may be cultivated in gardens year-round in subtropical climates and during the summer in temperate zones. They are popular houseplants and are commonly sold in floral shops and nurseries in the winter months.

Toxic Part: The whole plant is injurious.

Toxin: Raphides of water-insoluble calcium oxalate and unverified proteinaceous toxins.

Clinical Findings: A painful burning sensation of the lips and mouth results from ingestion. There is an inflammatory reaction, often with edema and blistering. Hoarseness, dysphonia, and dysphagia may result.

Caladium bicolor

Caladium bicolor

Management: The pain and edema recede slowly without therapy. Cool liquids or demulcents held in the mouth may bring some relief. Analgesics may be indicated. The insoluble oxalate in these plants does not cause systemic oxalate poisoning. Consultation with a Poison Control Center should be considered. See "Poisoning by Plants with Calcium Oxalate Crystals," p. 23.

References

Armien AG, Tokarnia CH. Experiments on the toxicity of some ornamental plants in sheep. Pesqui Vet Bras 1994;14:69–73.

Franceschi VR, Nakata PA. Calcium oxalate in plants: Formation and function. Annu Rev Plant Biol 2005;56:41–71.

Rauber A. Observations on the idioblasts of *Dieffenbachia*. J Toxicol Clin Toxicol 1985;23:79–90.

Schilling R, Marderosian A, Speaker J. Incidence of plant poisonings in Philadelphia noted as poison information calls. Vet Hum Toxicol 1980;22:148–150.

Calla palustris L.
Family: Araceae

Calla palustris

Common Names: Female Water Dragon, **Water Arum**, Water Dragon, Wild Calla

Description: This small plant has heart-shaped leaves that are 4 to 6 inches long and borne on 10-inch stalks. The spathe, or flower bract, is about 2 inches long, inconspicuous, green on the outside, and white on the inside. Red berries form on the spadix of the flower in thick clusters.

Distribution: This plant is found in wet, boggy areas from Quebec to Alberta to north central Alaska and south central Yukon south to Colorado, Texas, and Florida.

Toxic Part: The whole plant, particularly the root, is injurious.

Toxin: Raphides of water-insoluble calcium oxalate and unverified proteinaceous toxins.

Clinical Findings: A painful burning sensation of the lips and oral cavity results from ingestion. There is an inflammatory reaction, often with edema and blistering. Hoarseness, dysphonia, and dysphagia may result.

Management: The pain and edema recede slowly without therapy. Cool liquids or demulcents held in the mouth may bring some relief. Analgesics may be indicated. The insoluble oxalate in these plants does not cause systemic oxalate poisoning. Consultation with a Poison Control Center should be considered. See "Poisoning by Plants with Calcium Oxalate Crystals," p. 23.

References

Airaksinen MM, Peura P, Ala-Fossi-Salokangas L, et al. Toxicity of plant material used as emergency food during famines in Finland. J Ethnopharmacol 1986;18:273–296.

Rauber A. Observations on the idioblasts of *Dieffenbachia*. J Toxicol Clin Toxicol 1985;23:79–90.

Calophyllum inophyllum L.
Family: Guttiferae (Clusiaceae)

Common Names: Alexandrian Laurel, Beautyleaf, Indian Laurel, Kamani, Laurelwood, María Grande, **Mastwood**, Poonay Oil Plant

Description: This low-branched tree grows to 40 feet or more. The trunk is covered by light gray bark with a pink inner bark. The paired leaves are shiny green and leathery with a prominent midrib, 3 to 8 inches long and elliptical with a slight notch at the tip. The flowers form small clusters resembling orange blossoms and are fragrant. The fruit is globose, 1.25 to 1.5 inches in diameter, and becomes yellowish-brown at maturity. It contains one large seed with a bony shell. The plant has a cream-colored, resinous latex.

Calophyllum inophyllum, fruiting branch

Distribution: This tree is native to India and the Malay Peninsula. It is a commonly planted ornamental in the West Indies, south Florida, Hawaii, and Guam.

Toxic Part: The seed kernel is poisonous.

Calophyllum inophyllum, leaf and fruit (above)

Calophyllum inophyllum, flowers (left)

Toxin: Inophyllum A-E, calophyllolide, calophynic acid.

Clinical Findings: Ingestion can cause nausea, vomiting, abdominal cramping, diarrhea, and dehydration. The latex is used as a home remedy in the West Indies and the Pacific and appears to be nontoxic, although it may cause keratoconjunctivitis on contact with the cornea.

Management: Most exposures do not lead to significant toxicity. Intravenous hydration, antiemetics, and electrolyte replacement may be necessary in severe cases, particularly in children. Consultation with a Poison Control Center should be considered. See "Poisoning by Plants with Gastrointestinal Toxins," p. 28.

References

Holdsworth DK, So-On NC. Medicinal and poisonous plants from Manus Island. Sci New Guinea 1973;1:11–16.

Itoigawa M, Ito C, Tan HT, et al. Cancer chemopreventive agents, 4-phenylcoumarins from *Calophyllum inophyllum*. Cancer Lett 2001;169:15–19.

Ravelonjato B, Libot F, Ramiandrasoa F, Kunesch N, Gayral P, Poisson J. Molluscicidal constituents of *Calophyllum* from Madagascar: activity of some natural and synthetic neoflavonoids and khellactones. Planta Med 1992;58:51–55.

Calotropis species
Family: Asclepiadaceae
Calotropis gigantea (L.) W.T. Aiton
Calotropis procera (Aiton) W.T. Aiton

Common Names:
Calotropis gigantea: Bowstring Hemp, **Crown Flower**, Giant Milkweed, Mudar, Mudar Crown Plant, Pua Kalaunu
Calotropis procera: Algodón de Seda, French Jasmine, Giant Milkweed, Mudar, Mudar Small Crown Flower, **Small Crown Flower**, Tula

Description: These treelike shrubs have ovate or elliptical thick, glaucous, rubbery, opposite leaves. The flowers appear in clusters along the branches; they have a prominent crown with recurved petals and a sweet, pleasant odor. Colors vary from creamy white to lilac, mauve, and purple. The seeds have silky attachments (like other types of milkweed seeds), which emerge from pods as they split on drying. The two species differ in size: *Calotropis gigantea* grows to 15 feet; *C. procera* generally grows to under 6 feet and has correspondingly more diminutive plant parts.

Distribution: Crown flowers are cultivated in south Florida, California, and Hawaii. *Calotropis procera* is a weed in the West Indies.

Toxic Part: The latex has a direct irritant action on mucous membranes, particularly in the eye. Skin reactions to this plant may be caused by allergy rather than to a direct irritant action. All parts of the plant contain a cardioactive steroid and calcium oxalate crystals.

Calotropis gigantea, tip of branch with leaves and flowers

Toxin: An unidentified vesicant allergen in the latex, calcium oxalate crystals, and cardioactive steroids resembling digitalis.

Clinical Findings: Human intoxications from this plant have not been reported in modern times. Ingestion of calcium oxalates causes a painful burning sensation of the lips and mouth. There is an inflammatory reaction, often with edema and blistering. Hoarseness, dysphonia, and dysphagia may result.

Poisoning would be expected to produce clinical findings typical of cardioactive steroids. Toxicity has a variable latent period that depends on the quantity ingested. Dysrhythmias include sinus bradycardia, premature ventricular contractions, atrioventricular conduction defects, or ventricular tachydysrhythmias. Hyperkalemia, if present, may be an indicator of toxicity.

Management:
Calcium oxalate toxicity: The pain and edema recede slowly without therapy. Cool liquids or demulcents held in the mouth may bring some relief. Analgesics may be indicated. The insoluble oxalate in these plants does not cause systemic oxalate poisoning. See "Poisoning by Plants with Calcium Oxalate Crystals," p. 23.

Calotropis procera

Calotropis procera, tip of branch with leaves and flowers

Calotropis procera, mature fruits opening to release seeds

Cardioactive steroid toxicity: Gastrointestinal decontamination as appropriate, serial electrocardiograms, and serum potassium determinations should be performed. If serious cardioactive steroid toxicity is considered, digoxin-specific Fab should be administered. Consultation with a Poison Control Center should be considered. See "Poisoning by Plants with Cardioactive Steroids/Cardiac Glycosides," p. 24.

References

Handa F, Sadana JK, Sharma PK. Allergic contact dermatitis due to the plant *Calotropis procera*. A case report. Indian J Dermatol 1984;29:27–29.

Radford DJ, Gillies AD, Hinds JA, Duffy P. Naturally occurring cardiac glycosides. Med J Aust 1986;144:540–544.

Rutten AM, Statius van Eps LW. Poisoning with toxic plants in Curaçao in 1766. Ned Tijdschr Geneeskd 1998;142:2796–2798.

Caltha species
Family: Ranunculaceae
Caltha leptosepala DC.
Caltha palustris L.

Common Names: American Cowslip, Bull Flower, Cowslip, Gools, Horse Blob, Kingcup, **Marsh Marigold**, May, Meadow Bright, Meadow Gowan, Populage, Soldier's Buttons, Souci d'Eau, Water Goggles

Caltha palustris (right)

Caltha palustris, flowers (below)

Description: These plants are perennial herbs of the marshy wetlands. Their large, kidney-shaped leaves are about 7 inches wide. The stems are 8 to 24 inches long, smooth, hollow, and furrowed. *Caltha palustris* blooms in April or early May; the bright yellow flowers are about 2 inches in diameter and are followed by short, flat, many-seeded pods. *Caltha leptosepala* is an alpine bog plant that often blossoms at the lower edge of the retreating snow line in the spring.

Distribution: *Caltha palustris* grows in Alaska, Canada to the Arctic, and in the southeastern United States into North Carolina and Tennessee. *Caltha leptosepala* is found in Montana, Washington, British Columbia, Alberta, and Alaska. Other species, some with white flowers, also may be found in the northwestern states.

Toxic Part: All parts of the mature plant are poisonous. Before flowering, the immature plant may be boiled and eaten as greens.

Toxin: Protoanemonin, an irritant.

Clinical Findings: There are no adequately documented human poisonings, and clinical descriptions are derived primarily from animal reports. Intense pain and inflammation of the mouth with blistering, ulceration, and profuse salivation can occur. Bloody emesis and diarrhea develop in association with severe abdominal cramps. Central nervous system involvement is manifested by dizziness, syncope, and seizures.

Management: Most exposures result in minimal or no toxicity. Intravenous hydration, antiemetics, and electrolyte replacement may be necessary for patients with severe gastrointestinal symptoms, particularly in children. If seizures occur, rapidly acting anticonvulsants, such as intravenous diazepam, should be utilized along with other supportive measures. Consultation with a Poison Control Center should be strongly considered. See "Poisoning by Plants with Gastrointestinal Toxins," p. 28.

References

Bruni A, Bonora A, Dall'Olio G. Protoanemonin detection in *Caltha palustris*. J Nat Prod 1986;49:1172–1173.

Turner NJ. Counter-irritant and other medicinal uses of plants in Ranunculaceae by native peoples in British Columbia and neighbouring areas. J Ethnopharmacol 1984;11: 181–201.

Calycanthus species
Family: Calycanthaceae
Calycanthus floridus L.
(=*Calycanthus fertilis* Walter)
Calycanthus occidentalis Hook. & Arn.

Common Names: American Allspice, Bubbie, Bubby Blossoms, Bubby Bush, Calycanth, **Carolina Allspice**, Pineapple Shrub, Spicebush, Strawberry Bush, Strawberry Shrub, Sweet Shrub, Sweet Bettie

Description: *Calycanthus* grows as a shrub up to 12 feet in height. These plants have opposite leaves that are up to 6 inches long. The flowers are large (2 to 3 inches across), showy, and brownish-red or purple (rarely cream) with a pleasant, sweet, fragrant odor; they grow at the tips of small branchlets along the branches. The fruit is fig shaped and contains large glossy brown seeds.

Calycanthus floridus, flowering branch (above)

Calycanthus floridus, close-up of flower (below)

Distribution: *Calycanthus floridus* is native in the eastern United States from Pennsylvania to northern Florida and west to Alabama. *Calycanthus occidentalis* grows in California. These plants also are cultivated in parks in these areas.

Toxic Part: Seeds are poisonous.

Toxin: Calycanthin and related alkaloids. These toxins resemble the toxin strychnine.

Clinical Findings: There are no adequately documented human poisonings, and clinical descriptions are based on the nature of the toxin. Animals given the alkaloids experience strychnine-like convulsions, muscular hyperactivity, myocardial depression, and hypertension.

Management: If toxicity occurs, treatment entails aggressive management of muscle contractions with muscle relaxants and other supportive measures. Consultation with a Poison Control Center should be considered. See "Poisoning by Plants with Convulsive Poisons (Seizures)," p. 25.

Calycanthus floridus cv. 'Athens'

References

Bradley RE, Jones TJ. Strychnine-like toxicity of *Calycanthus*. Southeast Vet 1963;14: 40, 71, 73.

Collins RP, Halim AF. Essential leaf oils in *Calycanthus floridus*. Planta Med 1971;20:241–243.

Capsicum species
Family: Solanaceae
Capsicum annuum L. var. *annuum*
Capsicum annuum var. *glabriusculum* (Dunal) Heiser & Pickersgill
Capsicum chinense Jacq.
Capsicum frutescens L.

Common Names:
Capsicum annuum var. *annuum*: Ají de Gallina, Ají Guaguao, Cayenne Pepper, Cherry Pepper, **Chili Pepper**, Hot Pepper, Long Pepper, Nioi, Nioi-Pepa, Piment Bouc, Red Pepper, Sweet or Bell Pepper
Capsicum annuum var. *glabriusculum*: **Bird Pepper**
Capsicum chinense: **Habañero**, Scotch Bonnet, Squash Pepper
Capsicum frutescens: Ají Caballero, Ají Picante, Piment, **Tabasco Pepper**

Description: There are about 20 species of *Capsicum*. These perennial herbs are native to tropical America. The leaves are alternate, elliptical, and smooth edged. The flowers appear at the point of stem branching and are usually white with a purple tinge. The fruit is a pod containing many seeds; depending on the species and cultivar, the mature fruit can be variously colored—red, purple, orange, or yellow.

Toxic Part: The fruit and seeds of most peppers, with the exception of the sweet or bell pepper, are injurious.

Toxin: Capsaicin is a mucous membrane irritant that causes release of substance P from sensory nerve fibers. Substance P leads to stimulation of pain fibers and release of inflammatory mediators.

Capsicum annuum, fruit

Clinical Findings: Biting into a fruit leads to a burning or stinging sensation of the oral mucous membranes. Skin exposure may cause erythema but not vesiculation. Exposure to abraded skin can produce moderate to severe pain lasting several hours. (Paradoxically, capsaicin is available as a nonprescription medication that is applied to painful lesions to deplete the sensory nerve endings of substance P and provide analgesia.) Eye exposure to juice from this plant may lead to moderate chemical conjunctivitis and pain.

Management: Pain is managed by irrigation to remove remaining toxin and the administration of analgesics. Inflammatory reactions, particularly in the eye, may necessitate extensive irrigation with water and the use of topical analgesics.

References

Tominack RL, Spyker DA. Capsicum and capsaicin—a review: Case report of the use of hot peppers in child abuse. Clin Toxicol 1987;25:591–601.

Virus RM, Gebhart T. Pharmacologic actions of capsaicin: Apparent involvement of substance P and serotonin. Life Sci 1979;25:1273–1284.

Weinberg RB. Hunan hand. N Engl J Med 1981;31:1020.

Caryota species
Family: Palmae (Arecaceae)
Caryota mitis Lour.
Caryota urens L.

Common Names: Burmese Fishtail Palm, Cariota, Clustered Fishtail Palm, **Fishtail Palm**, Griffithii, Toddy Fishtail Palm, Tufted Fishtail Palm, Wine Palm

Description: These palms grow to 40 feet tall and have compound leaves up to 20 feet in length with small fishtail-shaped leaflets. The red or black fruits are about 0.5 inch in diameter and form on clusters of long, stringlike floral branches.

Caryota urens

Caryota gigas, leaf

Caryota sp., inflorescences on tree (left)

Caryota mitis, embryo showing calcium oxalate crystals (magnified ×20) (below)

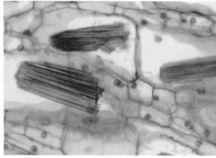

Distribution: *Caryota* are native to Asia, but are commonly grown in Florida, Guam, Hawaii, and the West Indies.

Toxic Part: The fruit pulp is injurious.

Toxin: Raphides of water-insoluble calcium oxalate and unverified proteinaceous toxins.

Clinical Findings: A painful burning sensation of the lips and oral cavity results from ingestion. There is an inflammatory reaction, often with edema and blistering. Hoarseness, dysphonia, and dysphagia may result.

Management: The pain and edema recede slowly without therapy. Cool liquids or demulcents held in the mouth may bring some relief. Analgesics may be indicated. The insoluble oxalate in these plants does not cause systemic oxalate poisoning. Consultation with a Poison Control Center should be considered. See "Poisoning by Plants with Calcium Oxalate Crystals," p. 23.

References

Rauber A. Observations on the idioblasts of *Dieffenbachia*. J Toxicol Clin Toxicol 1985; 23:79–90.

Snyder DS, Hatfield GM, Lampe KF. Examination of the itch response from the raphides of the fishtail palm *Caryota mitis*. Toxicol Appl Pharmacol 1979;48:287–292.

Cassia fistula L.
Family: Leguminosae (Fabaceae)

Common Names: Cañafístola, Casse, Golden Rain, **Golden Shower**, Indian Laburnum, Pudding-Pipe Tree, Purging Fistula

Cassia fistula, tree in flower

Description: The golden shower belongs to a genus with about 500 species of extremely showy, flowering ornamental trees and shrubs. This tree grows to 30 feet and has compound leaves that are 1 foot long with three to eight pairs of leaflets that are 2 to 6 inches in length. Large clusters of golden flowers in pendant racemes are followed by fruits that are long, narrow cylinders containing

Cassia fistula, close-up of flowers

Cassia fistula, branch showing flowers and seed pods

up to 100 flat seeds embedded in a sticky pulp in partitioned compartments.

Distribution: The golden shower is cultivated in subtropical areas, the West Indies, southern Florida, the coast of southern California, and Hawaii. Many related species of shrubs are native in the West Indies and have escaped from cultivation in Hawaii and Guam.

Toxic Part: The sticky fruit pulp is poisonous. The leaves and bark are less toxic.

Toxin: Emodin glycoside (senna), an anthraquinone cathartic.

Clinical Findings: Ingestion can cause nausea, vomiting, abdominal cramping, diarrhea, and dehydration. Emodin can also cause benign discoloration of the urine (yellowish-brown in acid urine, red or violet in basic urine).

Management: Intravenous hydration, antiemetics, and electrolyte replacement may be necessary for patients with severe gastrointestinal symptoms, particularly in children. Consultation with a Poison Control Center should be considered. See "Poisoning by Plants with Gastrointestinal Toxins," p. 28.

References

Barthakur NN, Arnold NP, Alli I. The Indian laburnum (*Cassia fistula* L.) fruit: An analysis of its chemical constituents. Plant Foods Hum Nutr 1995;47:55–62.

Mahesh VK, Sharma R, Singh RS, Upadhya SK. Anthraquinones and kaempferol from *Cassia* species section *fistula*. J Nat Prod 1984;47:733.

Cassia occidentalis L.
(=*Senna occidentalis* (L.) Link)
Family: Leguminosae (Fabaceae)

Common Names: 'Auko'I, Biche Prieto, Casse Puante, **Coffee Senna**, Hedionda, Mukipalaoa, Piss-a-Bed, Stinking Weed, Styptic Weed, Wild Coffee, Yerba Hedionda

Description: This annual herb grows to a height of 3 feet. The smooth leaves are pinnately compound with an even number of leaflets. The flowers are yellow but dry to a paler shade or white. The fruit is about 4.5 inches long and 0.25 inches wide. The seeds are dark olive-green.

Cassia occidentalis

Distribution: *Cassia occidentalis* is common in waste places, particularly along highways and coastal areas. It is probably native to the southeastern United States from Virginia to Texas and the West Indies but is now pantropical, including Hawaii and Guam.

Toxic Part: The whole plant is toxic; human intoxications usually involve raw seeds. The roasted seed may be used safely as a coffee substitute.

Toxin: The irritant chrysarobin (1,8-trihydroxy-3-methyl-9-anthrone), the cathartic emodin (1,8-trihydroxy-6-methyl-9,10-anthracenedione), and a lectin (toxalbumin) have been reported.

Clinical Findings: This plant is a major toxin in contaminated animal feed. Exposures in humans are rare. Ingestion of the raw seeds may exert gastrointestinal symptoms in humans. Chronic ingestion of seeds or other plant parts by animals leads to myodegeneration, muscle necrosis, and death.

Management: Intravenous hydration, antiemetics, and electrolyte replacement may be necessary for patients with severe gastrointestinal symptoms, particularly in children. Consultation with a Poison Control Center should be considered. See "Poisoning by Plants with Gastrointestinal Toxins," p. 28.

References

Calore EE, Cavaliere MJ, Haraguchi M, et al. Toxic peripheral neuropathy of chicks fed *Senna occidentalis* seeds. Ecotoxicol Environ Saf 1998;39:27–30.

Marrero Faz E, Bulnes Goicochea C, Perez Ruano M. *Cassia occidentalis* toxicosis in heifers. Vet Hum Toxicol 1998;40:307.

O'Hara PJ, Pierce KR, Reid WK. Degenerative myopathy associated with ingestion of *Cassia occidentalis*: clinical and pathologic features of the experimentally induced disease. Am J Vet Res 1969;30:2173–2180.

Suliman HB, Shommein AM. Toxic effect of the roasted and unroasted beans of *Cassia occidentalis* in goats. Vet Hum Toxicol 1986;28:6–11.

Catharanthus roseus (L.) G. Don
(=*Vinca rosea* L.)
Family: Apocynaceae

Catharanthus roseus

Common Names: Madagascar periwinkle, Bigleaf Periwinkle, Large Periwinkle, Periwinkle, Vinca (formerly known as *Vinca rosea*)

Description: The Madagascar periwinkle is a perennial herb with milky sap that is often cultivated as an annual. It has erect stems that bear dark glossy green, opposite, oblong-lanceolate leaves, 1 to 2 inches long, and bear solitary rose pink to white flowers about 1.5 inches across.

Distribution: Native from Madagascar to India, the plant has now spread throughout the tropics and is cultivated in many areas of the world as an ornamental

Catharanthus roseus, close-up of flower (above)

Catharanthus roseus, close-up of flowers (below)

garden plant. In temperate parts of the United States, it is cultivated as a summer annual.

Toxic Part: The whole plant is poisonous. A tea made from the leaves and stems is used in folk medicine in the Carribean and elsewhere.

Toxin: Vinca alkaloids (e.g., vincristine), clinically similar to colchicine, a cytotoxic alkaloid capable of inhibiting microtubule formation.

Clinical Findings: Ingestion may cause initial oropharyngeal pain followed in several hours by intense gastrointestinal symptoms. Abdominal pain and severe, profuse, persistent diarrhea may develop causing extensive fluid depletion and its sequelae. Vinca alkaloids may subsequently produce peripheral neuropathy, bone marrow suppression, and cardiovascular collapse.

Management: Aggressive symptomatic and supportive care is critical, with prolonged observation of symptomatic patients. Consultation with a Poison Control Center should be strongly considered. See "Poisoning by Plants with Mitotic Inhibitors," p. 29.

References

Duffin J. Poisoning the spindle: serendipity and discovery of the anti-tumor properties of the Vinca alkaloids. Pharm Hist 2002;44:64–76.

Wu ML, Deng JF, Wu JC, Fan FS, Yang CF. Severe bone marrow depression induced by an anticancer herb *Catharanthus roseus*. J Toxicol Clin Toxicol 2004;42:667–671.

Caulophyllum thalictroides (L.) Michx.
Family: Berberidaceae

Common Names: Blueberry Root, **Blue Cohosh**, Blue Ginseng, Papoose Root, Squaw Root, Yellow Ginseng

Description: The blue cohosh is an erect herb, 1 to 3 feet tall. The leaflets are lobed above the middle. The flowers are yellowish-green or greenish-purple, about 0.5 inch across, and occur in small clusters. The seeds are globose with a thin, fleshy blue coat.

Distribution: This plant is found in moist woody areas in Canada from southeast Manitoba to New Brunswick south to Alabama and west to Missouri.

Toxic Part: The seeds and roots are cytotoxic.

Toxin: Saponins and *N*-methylcytisine, a nicotine-like alkaloid.

Clinical Findings: The extremely bitter taste usually limits the quantity ingested. The saponins are gastrointestinal irritants that can cause nausea, vomiting, abdominal cramping, diarrhea, and dehydration. The *N*-methylcytisine causes nicotine-like clinical effects. Initial gastrointestinal effects may be followed by those typical of nicotine poisoning; these include hypertension, large pupils, sweating, and perhaps seizures. Severe poisoning produces coma, weakness, and paralysis that may result in death from respiratory failure.

Caulophyllum thalictroides, leaves (left)

Caulophyllum thalictroides, flowers (below)

Management: If severe gastrointestinal symptoms occur, intravenous hydration, antiemetics, and electrolyte replacement may be necessary, particularly in children. See "Poisoning by Plants with Gastrointestinal Toxins," p. 28. Symptomatic and supportive care should be given, with attention to adequacy of ventilation and vital signs. Atropine may reverse some of the toxic effects. Consultation with a Poison Control Center should be strongly considered. See "Poisoning by Plants with Nicotine-like Alkaloids," p. 30.

References

Betz JM, Andrzejewski D, Troy A, et al. Gas chromatographic determination of toxic quinolizidine alkaloids in blue cohosh (*Caulophyllum thalictroides* (L.) Michx.). Phytochem Anal 1998;9:232–236.

Jones TK, Lawson BM. Profound neonatal congestive heart failure caused by maternal consumption of blue cohosh herbal medication. J Pediatr 1998;132:550–552.

Rao RB, Hoffman RS. Nicotinic foxicity from tincture of blue cohosh (Caulophyllum thalictroides) used as an abortifacient. Vet Hum Toxicol 2002;44:221–222.

Celastrus orbiculatus Thunb.
Celastrus scandens L.
Family: Celastraceae

Common Names:
Celastrus orbiculatus: **Oriental Bittersweet**
Celastrus scandens: American Bittersweet, **Bittersweet**, Bourreau des Arbres, Climbing Bittersweet, Climbing Orange Root, False Bittersweet, Fever Twig, Frobusk, Red Root, Roxbury Waxwork, Shrubby Bittersweet, Staff Tree, Staff Vine

Celastrus scandens, fruiting branch

Description: The bittersweet is a climbing, vinelike shrub that may grow to a height of 25 feet. Its ovate, toothed, alternate leaves are about 4 inches long. The flowers are inconspicuous. The mature fruit is a 0.5-inch, orange-yellow capsule, which opens to reveal a crimson pulp surrounding each seed.

Distribution: *Celastrus* grows in southeastern Saskatchewan to Quebec, south to North Carolina, and west to New Mexico. It is much employed for dried arrangements in the fall and is readily available from florists at that time. *Celastrus orbiculatus* is a native of eastern Asia and now has become an abundant invasive vine in areas of the United States. *Celastrus scandens* is native to North America.

Celastrus scandens, flowers (above)

Toxic Part: The fruit is reputed to be poisonous; all parts of this plant may be toxic.

Celastrus scandens (left) and *C. orbiculatus* (right), showing mature fruit (below)

Toxin: An unknown gastro-intestinal irritant.

Clinical Findings: Exposures are rare. Ingestion may cause nausea, vomiting, abdominal cramping, diarrhea, and dehydration.

Management: Intravenous hydration, antiemetics, and electrolyte replacement may be necessary for patients with severe gastrointestinal symptoms, particularly in children. Consultation with a Poison Control Center should be considered. See "Poisoning by Plants with Gastrointestinal Toxins," p. 28.

Reference

Rzadkowska-Bodalska H. Isolation and identification of flavonoids from leaves of *Celastrus scandens* L., *C. rugosa* Rehd. et Wills., *C. gemmata* Loes. and *C. loeseneri* Rehd. et Wills. Pol J Pharmacol Pharm 1973;25:407–416.

***Cestrum* species**
Family: Solanaceae
Cestrum diurnum L.
Cestrum nocturnum L.

Common Names:
Cestrum diurnum: Chinese Inkberry, Dama de Día, **Day Blooming Jessamine**, Day Blooming Cestrum, Day Jessmine, Galán de Día, Makahala
Cestrum nocturnum: 'Ala-Aumoe, Chinese Inkberry, Dama de Noche, Galán de Noche, Huele de Noche, Jasmín de Nuit, Kupaoa, Lilas de Nuit, **Night Blooming Jessamine**, Night Blooming Cestrum, Onaona-Iapana

Description: The day-blooming jessamine is an evergreen shrub or tall bush with smooth-edged ovate leaves about 2 to 4 inches long. The tubular flowers are white and appear in small clusters. They are fragrant by day. The mature fruit is a globose, black berry. The night-blooming jessamine has longer leaves (4 to 8 inches). The flowers are very fragrant at night and have a sweet odor. The mature fruit is white.

Cestrum diurnum, flowers

Distribution: Both species are native to the West Indies. They have escaped from cultivation in Florida, Texas, and Guam.

Toxic Part: Leaves.

Toxin: Calcinogenic glycosides of vitamin D (1,25-dihydroxy vitamin D_3) are found in *Cestrum diurnum*; *Cestrum nocturnum* contains atropine-like anticholinergic alkaloids.

Clinical Findings: There are no adequately documented human poisonings, and clinical descriptions are derived primarily from animal reports. Consumption of calcinogenic glycosides by grazing animals has caused vitamin D intoxication, leading to hypercalcemia, hypophosphatemia, bone softening, and deposition of calcium in the soft tissues (calcinosis), such as the kidney (nephrocalcinosis).

Anticholinergic alkaloid intoxication results in dry mouth with dysphagia and dysphonia, tachycardia, and urinary retention. Elevation of body temperature may be accompanied by flushed, dry skin. Mydriasis, blurred vision, excitement and delirium, headache, and confusion may be observed.

Cestrum nocturnum, flowers (above)

Cestrum nocturnum, fruiting branch (below)

Management: Intravenous hydration, antiemetics, and electrolyte correction may be necessary. If the severity of the anticholinergic alkaloid intoxication warrants intervention (hyperthermia, delirium), an antidote, physostigmine, is available. Consultation with a Poison Control Center should be considered for either syndrome. See "Poisoning by Plants with Anticholinergic (Antimuscarinic) Poisons," p. 21.

References

Halim AF, Collins RP, Berigari MS. Alkaloids produced by *Cestrum nocturnum* and *Cestrum diurnum*. Planta Med 1971;20:44–53.

Kasali OB, Krook L, Pond WG, Wasserman RH. *Cestrum diurnum* intoxication in normal and hyperparathyroid pigs. Cornell Vet 1977;67:190–221.

Sarkar K, Narbaitz R, Pokrupa R, Uhthoff HK. The ultrastructure of nephrocalcinosis induced in chicks by *Cestrum diurnum* leaves. Vet Pathol 1981;18:62–70.

Chelidonium majus L.
Family: Papaveraceae

Common Names: **Celandine**, Celandine Poppy, Elon Wort, Felonwort, Rock Poppy, Swallow Wort, Tetterwort

Description: This herb grows 1 to 3 feet in height. The yellow flower is 0.5 to 0.75 inch in diameter with four petals. Leaves are divided into three to five lobes. The leaves, stems, and buds are whitish and hairy. The sap is reddish-orange. This plant blooms from March to August.

Toxic Part: The whole plant is poisonous.

Toxin: Chelidonine, an isoquinoline alkaloid.

Clinical Findings: This plant is seldom ingested because of its unpleasant smell and taste. However, it is used as an herbal preparation for assorted maladies. Acute ingestions may cause nausea, vomiting, abdominal cramping, diarrhea, and dehydration. Chronic ingestions of herbals or extracts containing chelidonium may cause cholestatic liver toxicity.

Management: For acute toxicity, intravenous hydration, antiemetics, and electrolyte replacement may be necessary for patients with severe gastrointestinal symptoms, particularly children. Consultation with a Poison Control Center should be considered. See "Poisoning by Plants with Gastrointestinal Toxins," p. 28.

Chelidonium majus (left)

Chelidonium majus, close-up of leaves and flowers (below)

References

Benninger J, Schneider HT, Schuppan D, Kirchnor T, Hahn EG. Acute hepatitis induced by greater celandine (*Chelidonium majus*). Gastroenterology 1999;117:1234–1237.

Greving I, Meister V, Monnerjahn C, Muller KM, May B. *Chelidonium majus*: Rare reason for severe hepatotoxic reaction. Drug Saf 1998;7(suppl 1):S66–S69.

Spegazzini E, Najera M, Carpano S, Bergamini L, Esposito P. Valuation of chelidonine in *Chelidonium majus* L. (Papaveraceae) cultivated in Argentina. Acta Farm Bonaerense 1993; 12:69–72.

Chrysanthemum species
Family: Compositae (Asteraceae)

Common Names: Chrysanthemum, Mum

Description: The chrysanthemum is a very commonly cultivated plant consisting of 100 to 200 species of annuals or perennials, with innumerable hybrids introduced into the horticultural trade. They can be low growing and shrubby or 3 feet or more in height, depending on the species and the way it is grown. Leaves are alternate, sometimes forming a basal rosette, and often aromatic. The flowers are solitary or in clusters, and are white, yellow, orange, pink, red, or purple, and very showy.

Chrysanthemum sp., flower (above)

Chrysanthemum sp. (below)

Distribution: Native mostly to Europe and Asia with some from South Africa, these plants are very common in annual and perennial gardens and borders, as pot plants, and in the floral trade.

Toxic Part: Flower head.

Chrysanthemum sp., flower

Toxin: Arteglasin A, and other sesquiterpene lactones.

Clinical Findings: Contact dermatitis caused by contact sensitization is a common complaint among handlers of this plant. Immediate hypersensitivity (allergic) reactions are common.

Management: Symptomatic and supportive care. Consultation with a Poison Control Center should be considered.

References

de Jong NW, Vermeulen AM, van Wijk RG, de Groot H. Occupational allergy caused by flowers. Allergy 1998;53:204–209.

Kuno Y, Kawabe Y, Sakakibara S. Allergic contact dermatitis associated with photosensitivity, from alantolactone in a chrysanthemum farmer. Contact Dermatitis 1999;40:224–225.

Cicuta species
Family: Umbelliferae (Apiaceae)
Cicuta bulbifera L.
Cicuta douglasii J. M. Coult. & Rose
Cicuta maculata L.

Common Names: Water Hemlock, Beaver Poison, Children's Bane, Cicutaire, Death-of-Man, Musquash Poison, Musquash Root, Spotted Cowbane

Description: Most species of *Cicuta* have a similar appearance and may grow to a height of 6 feet. The leaves are two or three times pinnately compound. The flowers are small, whitish, and heavily scented. In the underground portion of mature plants, there is a bundle of chambered tuberous roots. A yellow, oily sap emerges from the cut stem and has a distinct odor of raw parsnip. These plants are found only in wet or swampy ground.

Distribution: Species of *Cicuta* are found throughout the United States (except Hawaii) and Canada and are among the most toxic plants in North America.

Toxic Part: The whole plant, particularly the roots, is poisonous.

Toxin: Cicutoxin, an unsaturated, long-chain aliphatic alcohol, a gamma-aminobutyric acid (GABA) receptor antagonist or potassium channel blocker.

Clinical Findings: Toxic ingestions often occur when the plant is mistaken for a similar appearing but edible wild plant such as *Daucus carota* L. (Queen Anne's lace). Onset of symptoms is rapid, usually within 1 hour of ingestion. Symptoms include nausea, vomiting, salivation, and trismus. Generalized seizures also may occur. Death may occur if seizures do not terminate.

Management: Supportive care, including airway management and protection against rhabdomyolysis-associated com-

Cicuta maculata

plications (e.g., electrolyte abnormalities and renal insufficiency) should be given. Rapidly acting anticonvulsants, such as diazepam or lorazepam, should be administered for persistent seizures. Consultation with a Poison Control Center should be considered. See "Poisoning by Plants with Convulsive Poisons (Seizures)," p. 25.

References

Carlton BE, Tufts E, Girard DE. Water hemlock poisoning complicated by rhabdomyolysis and renal failure. Clin Toxicol 1979;14:87–92.

CDC. Water hemlock poisoning—Maine 1992. MMWR (Morb Mortal Wkly Rep) 1994;43: 229–331.

Heath KB. A fatal case of apparent water hemlock poisoning. Vet Hum Toxicol 2001;43: 35–36.

Knutsen OH, Paszkowshi P. New aspects in the treatment of water hemlock poisoning. J Toxicol Clin Toxicol 1984;122:157–166.

Straub U, Wittstock U, Schubert R, et al. Cicutoxin from *Cicuta virosa*—a new and potent potassium channel blocker in T lymphocytes. Biochem Biophys Res Commun 1996;219: 332–336.

Clematis species
Family: Ranunculaceae

Common Names: Clematis, Blue Jessamine, Bluebell, Cabeza de Viejo, Cascarita, Clematite, Curly Heads, Devil's Hair, Devil's Thread, Flamula, Headache Weed, Leather Flower, Herbe aux Geaux, Curl Flower, Liane Bon Garçon, Pipe-stem, Sugar-bowls, Traveler's Joy, Vase Flower, Vase Vine, Virgin's Bower, Yerba de Pordioseros.

Description: There are many species of this perennial flowering herb and climber. The leaves are usually compound and have a superficial resemblance to poison ivy. The flowers can range from small in the native species to large and very showy in the hybrid cultivars. The fruits are small, flattened, and terminated by a plume of white hairs.

Clematis cv. 'Jackmanii,' close-up of flower (above)

Clematis cv. 'Niobe' (below)

Distribution: *Clematis* are native to Canada and the north temperate United States; cultivated varieties are available commercially throughout the United States and West Indies and are commonly used as vines in gardens.

Toxic Part: The whole plant is toxic.

Toxin: Protoanemonin, an irritant.

Clinical Findings: There are no adequately documented human poisonings, and clinical descriptions are derived primarily from animal reports. Intense pain and inflammation of the mouth with blistering, ulceration, and profuse salivation can occur. Bloody emesis and diarrhea develop in association with severe abdominal

Clematis paniculata

Clematis virginiana

cramps. Central nervous system involvement is manifested by dizziness, syncope, and seizures.

Management: Most exposures result in minimal or no toxicity. Intravenous hydration, antiemetics, and electrolyte replacement may be necessary for patients with severe gastrointestinal symptoms, particularly in children. Consultation with a Poison Control Center should be considered. See "Poisoning by Plants with Gastrointestinal Toxins," p. 28.

If seizures occur, rapidly acting anticonvulsants, such as intravenous diazepam, should be utilized along with other supportive measures. Consultation with a Poison Control Center should be strongly considered.

References

Cos P, Hermans N, De Bruyne T, et al. Further evaluation of Rwandan medicinal plant extracts for their antimicrobial and antiviral activities. J Ethnopharmacol 2002;79:155–163.

Kern JR, Cardellina JH. Native American medicinal plants. Anemonin from the horse stimulant *Clematis hirsutissima*. J Ethnopharmacol 1983 Jul;8:121–123.

Lancaster AH. *Clematis* dermatitis. South Med J 1937;30:207.

Shao B, Qin G, Xu R, Wu H, Ma K. Triterpenoid saponins from *Clematis chinensis*. Phytochemistry 1995;38:1473–1479.

Turner NJ. Counter-irritant and other medicinal uses of plants in Ranunculaceae by native peoples in British Columbia and neighbouring areas. J Ethnopharmacol 1984;11:181–201.

Clivia **species**
Family: Amaryllidaceae

Common Name: Kaffir Lily

Description: Straplike leaves arise from a swollen leaf base (bulb) and may be arranged in two opposing rows. The showy orange or red flowers form in a cluster on a leafless stalk. The mature fruit is a red, pulpy berry.

Clivia miniata

Distribution: Although native to Africa, this lily is commonly cultivated as a house or garden plant.

Toxic Part: All parts of this plant are toxic.

Toxin: Lycorine and related phenanthridine alkaloids (see *Narcissus*).

Clinical Findings: Ingestion of small amounts produces few or no symptoms. Large exposures may cause nausea,

Clivia nobilis (right)

Clivia nobilis, flowers (below)

vomiting, abdominal cramping, diarrhea, dehydration, and electrolyte imbalance.

Management: Intravenous hydration, antiemetics, and electrolyte replacement may be necessary for patients with severe gastrointestinal effects, particularly in children. Consultation with a Poison Control Center should be considered. See "Poisoning by Plants with Gastrointestinal Toxins," p. 28.

References

Amico A, Stefanizzi L, Bruno S. Osservazioni morfologiche, estrazione e localizzazione di alcaloidi in *Clivia miniata* Regel. Fitoterapia 1979;50:157.

Morton JF. *Plants Poisonous to People in Florida and Other Warm Areas.* Hurricane Press, Miami, 1971.

Clusia rosea Jacq.
Family: Guttiferae (Clusiaceae)

Common Names: Autograph tree, **Balsam Apple**, Copey, Cupey, Figuier Maudit Marron, Pitch Apple, Scotch Attorney, Wild Mamee

Description: This tree may be 20 to 50 feet high and grows on rocks or other trees. The leaves are ovate and leathery, 3 to 8 inches in diameter. The flowers are white tinged with pink and have a golden center. Globulose fruit, the size of a golf ball, turn brown and open when mature.

Distribution: Species of *Clusia* grow in the West Indies, southern Florida, and Hawaii.

Toxic Part: The golden viscous sap and the fruit are toxic.

Toxin: An unidentified gastrointestinal irritant (possibly nemorosone).

Clusia rosea, close-up of flower

Clinical Findings: Nausea, vomiting, abdominal cramping, and diarrhea may occur.

Management: If severe gastrointestinal symptoms occur, intravenous hydration, antiemetics, and electrolyte replacement may be necessary, particularly in children. Consultation with a Poison Control Center

Clusia rosea, open fruit (above)

Clusia rosea, fruit (right)

should be considered. See "Poisoning by Plants with Gastrointestinal Toxins," p. 28.

Reference

Cuesta-Rubio O, Velez-Castro H, Frontana-Uribe BA, Cardenas J. Nemorosone, the major constituent of floral resins of *Clusia rosea*. Phytochemistry 2001;57:279–283.

Colchicum species
Bulbocodium species
Family: Liliaceae
Bulbocodium vernum L. (=*Colchicum vernum* (L.) Ker-Gawl.)
Colchicum autumnale L.
Colchicum speciosum Steven

Common Names: Autumn Crocus, Crocus, Fall Crocus, **Meadow Saffron**, Mysteria, Naked Boys, Vellorita, Wonder Bulb

Description: *Colchicum* species are members of the Lily family and are cultivated for their long, tubular, purple or white flowers, which emerge from the underground bulb. *Colchicum autumnale* is a species that produces flowers in the following autumn, which then die back, and its leaves appear only during the spring.

Colchicum autumnale, flowers in Autumn (above)

Colchicum autumnale, flowers in Autumn (left)

Distribution: *Colchicum* species are grown outdoors in gardens or may be forced into bloom in pots.

Toxic Part: The whole plant is poisonous. *(Colchicum autumnale* is a commercial source for medicinal colchicine.)*

Toxin: Colchicine, a cytotoxic alkaloid capable of inhibiting microtubule formation.

Colchicum autumnale, leaves appearing in Spring

Clinical Findings: Ingestion may cause initial oropharyngeal pain followed in several hours by intense gastrointestinal symptoms. Abdominal pain and severe, profuse, persistent diarrhea may develop, causing extensive fluid depletion and its sequelae. Colchicine may subsequently produce peripheral neuropathy, bone marrow suppression, and cardiovascular collapse.

Management: The intoxication has a prolonged course due to the slow absorption of colchicine. Aggressive symptomatic and supportive care is critical, with prolonged observation of symptomatic patients. Consultation with a Poison

Control Center should be strongly considered. See "Poisoning by Plants with Mitotic Inhibitors," p. 29.

References

Brvar M, Kozelj G, Osredkar J, Mozina M, Gricar M, Bunc M. Acute poisoning with autumn crocus (*Colchicum autumnale* L.). Wien Klin Wochenschr. 2004;116:205–208.

Danel VC, Wiart JF, Hardy GA, Vincent FH, Houdret NM. Self-poisoning with *Colchicum autumnale* L. flowers. J Toxicol Clin Toxicol 2001;39:409–411.

Harris R, Marx G, Gillett M, Kark A, Arunanthy S. Colchicine-induced bone marrow suppression: Treatment with granulocyte colony-stimulating factor. J Emerg Med 2000;18:435–440.

Klintschar M, Beham-Schmidt C, Radner H, Henning G, Roll P. Colchicine poisoning by accidental ingestion of meadow saffron (*Colchicum autumnale*): Pathological and medicolegal aspects. Forensic Sci Int 1999;106:191–200.

Mullins ME, Carrico EA, Horowitz BZ. Fatal cardiovascular collapse following acute colchicine ingestion. J Toxicol Clin Toxicol 2000;38:51–54.

Colocasia species
Family: Araceae
Colocasia esculenta (L.) Schott (=*C. esculenta* (L.) Schott var. *antiquorum* (Schott)
     F. T. Hubb. & Rehder; *C. antiquorum* Schott)
Colocasia gigantea (Blume) Hook. f.

Common Names: Caraibe, Cocoyam, Dasheen, Eddo, **Elephant's Ear**, Kalo, Malanga Islena, Malanga deux Palles, **Taro**, Tayo Bambou, Yautía Malanga

Colocasia esculenta (right)

Colocasia esculenta (below)

Description: These large ornamentals are grown for their huge, shiny, heart-shaped leaves or for their edible tuber. The flowers are in a short spadix surrounded by a yellow spathe.

Distribution: Elephant's ear grows in the West Indies, south Florida, southern California, Guam, and Hawaii as landscape plants or for food. They are sometimes used as indoor plants in the remainder of the country.

Toxic Part: The leaves are injurious.

Toxin: Insoluble crystals (raphides) of calcium oxalate.

Clinical Findings: A painful burning sensation of the lips and mouth result from ingestion. There is an inflammatory reaction, often with edema and blistering. Hoarseness, dysphonia, and dysphagia may result.

Management: The pain and edema recede slowly without therapy. Cool liquids or demulcents held in the mouth may bring some relief. Analgesics may be indicated. The insoluble oxalate in these plants does not cause systemic oxalate poisoning. Consultation with a Poison Control Center should be considered. See "Poisoning by Plants with Calcium Oxalate Crystals," p. 23.

References

Mihailidou H, Galanakis E, Paspalaki P, Borgia P, Mantzouranis E. Pica and the elephant's ear. J Child Neurol 2002;17:855–856.

Tagwireyi D, Ball DE. The management of Elephant's Ear poisoning. Hum Exp Toxicol 2001;20:189–192.

Conium maculatum **L.**
Family: Umbelliferae (Apiaceae)

Common Names: Bunk, California Fern, Cashes, Cigue, Herb Bonnett, Kill Cow, Nebraska Fern, **Poison Hemlock**, Poison Parsley, Poison Root, St. Bennett's Herb, Snake Weed, Spotted Hemlock, Spotted Parsley, Winter Fern, Wode Whistle

Description: This plant resembles wild carrot (*Daucus carota* L.) with large lacy leaves, but the root is white. The leaves may extend to 4 feet, and both stem and leaves may be spotted with purple. When crushed, the leaves produce an offensive "mousy" odor. The small white flowers are borne in flat clusters (umbels).

Distribution: This plant is now naturalized in most of the northern and temperate zones of the United States and Canada (but not Alaska).

Toxic Part: The whole plant is poisonous, particularly the root and seeds.

Conium maculatum, close-up of flowers (above)

Conium maculatum (right)

Conium maculatum, leaves

Toxin: Coniine and related nicotine-like alkaloids.

Clinical Findings: Given its resemblance to wild carrot, misidentification accounts for most poisonings. Initial gastrointestinal effects may be followed by those typical of nicotine poisoning; these include hypertension, large pupils, sweating, and perhaps seizures. Severe poisoning produces coma, weakness, and paralysis that may result in death from respiratory failure.

Management: Symptomatic and supportive care should be given, with attention to adequacy of ventilation and vital signs. Atropine may reverse some of the toxic effects. Consultation with a Poison Control Center should be strongly considered. See "Poisoning by Plants with Nicotine-like Alkaloids," p. 30.

References

Drummer OH, Roberts AN, Bedford PJ, Crump KL, Phelan MH. Three deaths from hemlock poisoning. Med J Aust 1995;162:592–593.

Frank BS, Michelson WB, Panter KE, Gardner DR. Ingestion of poison hemlock (*Conium maculatum*). West J Med 1995;163:573–574.

Lopez TA, Cid MS, Bianchini ML. Biochemistry of hemlock (*Conium maculatum* L.) alkaloids and their acute and chronic toxicity in livestock. A review. Toxicon 1999;37: 841–865.

Vetter J. Poison hemlock (*Conium maculatum* L.). Food Chem Toxicol 2004;42:1373–1382.

Convallaria majalis L.
Family: Liliaceae

Common Names: Conval Lily, Lilia-O-Ke-Awawa, **Lily-of-the-Valley**, Mayflower, Muguet

Description: These small perennials have two oblong leaves and a flower stalk bearing small, drooping, bell-shaped white flowers on one side. Occasionally, orange-red fleshy berries form. *Convallaria* spreads by underground roots to form thick beds.

Distribution: These Eurasian plants have escaped from cultivation throughout the northern and temperate United States and eastern Canada. They are

Convallaria majalis, individual plant in flower (left)

Convallaria majalis, used as ground cover (below)

commonly sold as potted plants in the florist trade, especially during the winter months when they are forced into flower, as well as used in landscaping.

Toxic Part: The whole plant is poisonous, as is the water in which flowers have been kept.

Toxin: Convallotoxin, a digitalis-like cardioactive steroid, and irritant saponins.

Clinical Findings: Most exposures result in minimal or no toxicity. Large exposures have the potential to cause systemic toxicity. Pain in the oral cavity, nausea, emesis, abdominal pain, cramping, and diarrhea develop after ingestion. Dysrhythmias have a variable latent period that depends on the quantity ingested. Symptoms are usually expressed as sinus bradycardia, premature ventricular contractions, atrioventricular conduction defects, or ventricular tachydysrhythmias. Hyperkalemia, if present, may be an indicator of toxicity.

Management: Gastrointestinal decontamination as appropriate, serial electrocardiograms, and serum potassium determinations should be performed. If poisoning is seriously considered, digoxin-specific Fab should be administered. Consultation with a Poison Control Center should be considered. See "Poisoning by Plants with Cardioactive Steroids/Cardiac Glycosides," p. 24.

References

Edgerton PH. Symptoms of digitalis-like toxicity in a family after accidental ingestion of lily-of-the-valley plant. J Emerg Nurs 1989;15:220–223.

Krenzelok EP, Jacobsen TD, Aronis J. Lily-of-the-valley (*Convallaria majalis*) exposures: Are the outcomes consistent with the reputation? J Toxicol Clin Toxicol 1996;34:601.

Lamminpaa A, Kinos M. Plant poisonings in children. Hum Exp Toxicol 1996;15:245–249.

Moxley RA, Schneider NR, Steinegger DH, Carlson MP. Apparent toxicosis associated with lily-of-the-valley (*Convallaria majalis*) ingestion in a dog. J Am Vet Med Assoc 1989;195:485–487.

Coriaria myrtifolia L.
Family: Coriariaceae

Common Names: Myrtle-leaved Sumac, Myrtle-leaved Tanner's Sumac, Redoul

Description: Species of *Coriaria* are shrubs or small trees with opposite leaves, small greenish flowers, and a purple-black berry-like fruit.

Distribution: *Coriaria* are in cultivation as an ornamental shrub in the southern United States and California.

Toxic Part: The fruit is poisonous.

Toxin: Coriamyrtin, a gamma-aminobutyric acid (GABA) antagonist.

Clinical Findings: Exposures often involve ingestion of the berries. Most cases have few or no symptoms. Substantial ingestion causes excitatory neurological effects including seizures.

Coriaria myrtifolia, close-up of leaves and fruit (above)

Coriaria myrtifolia (left)

Management: Gastrointestinal decontamination. If seizures occur, rapidly acting anticonvulsants, such as intravenous diazepam, should be utilized along with other supportive measures. Consultation with a Poison Control Center should be considered. See "Poisoning by Plants with Convulsive Poisons (Seizures)," p. 25.

References

Alonso Castell P, Moreno Galdo A, Sospedra Martínez E, Rogueta Mast, Hidalgo Albert E, Iglesias Birengve J. Serious intoxication by *Coriaria myrtifolia*: A case study. An Esp Pediatr 1997;46:81–82.

de Haro L, Pommier P, Tichadou L, Hayek-Lanthois M, Arditti J. Poisoning by *Coriaria myrtifolia* Linnaeus: a new case report and review of the literature Toxicon 2005;64:600–603.

Skalli S, David JM, Benkirane R, Zaid A, Soulaymani R. Acute intoxication by redoul (*Coriaria myrtifolia* L.). Three observations. Presse Med 2002;31:1554–1556.

Corynocarpus laevigatus J.R. Forst. & G. Forst.
Family: Corynocarpaceae

Common Names: This species is referred to as Karaka nut in New Zealand.

Description: *Corynocarpus* is a tree with thick, dark green, glossy leaves. The greenish-yellow flowers are borne in panicles. The plum-shaped fruit is green, turning to orange when mature, and 1 to 1.5 inches long.

Corynocarpus laevigatus, close-up of flower buds (above)

Corynocarpus laevigatus (left)

Distribution: *Corynocarpus* is native to New Zealand and has been introduced in California, Mexico, and the Gulf Coast.

Toxic Part: The uncooked seeds are poisonous, but the ripe flesh of the fruit is edible.

Toxin: Hydrolysis of the prototoxin karakin yields the toxin beta-nitropropionic acid, an irreversible inhibitor of succinate dehydrogenase, a mitochondrial Krebs cycle enzyme.

Clinical Findings: There are no adequately documented human poisonings, and clinical descriptions are derived primarily from animal reports, in which admininstration produces convulsions followed by cardiovascular collapse.

Management: Symptomatic and supportive care should be given. Consultation with a Poison Control Center should be considered. See "Poisoning by Plants with Convulsive Poisons (Seizures)," p. 25.

Reference

Bell ME. Toxicology of karaka kernel, karakin, and beta-nitropropionic acid. N Z J Sci 1974;17:327–334.

Crassula argentea Thunb.
Family: Crassulaceae

Common Names: Baby Jade, Cauliflower Ears, Chinese Rubber Plant, Dollar Plant, Dwarf Rubber Plant, **Jade Plant**, Jade Tree, Japanese Rubber Plant

Description: This is a small pot plant that can grow as a shrub to 10 feet tall. The leaves are obovate in shape, thick and fleshy, 1 to 2 inches long, and sometimes tinged in red along the margins. Some cultivars of this species are variegated with white, blue, and cream. The plant has white or pink flowers when planted outdoors but rarely flowers when kept as a pot plant indoors in cooler climates.

Distribution: A plant originally from Southern Africa, it is grown indoors as an ornamental where it is common, and outdoors in warmer areas that do not go below 40°F.

Toxic Part: It is unclear what toxin is responsible for toxicity when it occurs.

Toxin: Unknown.

Clinical Findings: This plant is commonly ingested by children because of its wide availability as a household plant. Typical ingestions do not lead to systemic toxicity. Exposure to the sap from the plant or leaves may produce contact dermatitis.

Management: Symptomatic and supportive care. Consultation with a Poison Control Center should be considered.

Crassula argentea (left)

Crassula argentea, close-up of leaves and flowers (below)

Reference

Krenzelok EP, Jacobsen TD. Plant exposures: A national profile of the most common plant genera. Vet Hum Tox 1997;39:248–249.

Crinum species
Family: Amaryllidaceae
Crinum asiaticum L.
Crinum bulbispermum (Burm.) Milne-Redh. & Schweick.

Common Names: Crinum Lily, Lirio, Lys, **Spider Lily**

Description: These plants have onion-like bulbs. The straplike leaves are arranged in a spiral. Lily-like flowers appear on a solid, leafless stem and are usually white but may be pink or red.

Distribution: Some species are native to the southern United States and the West Indies. *Crinum bulbispermum*, from South Africa, and *C. asiaticum* L., from tropical Asia, are most frequently encountered in cultivation.

Toxic Part: All parts of the plant are toxic, particularly the bulb.

Toxin: Lycorine and related phenanthridine alkaloids (see *Narcissus*).

Clinical Findings: Toxicity is uncommonly reported in humans. Ingestion of small amounts produces few or no symptoms. Large exposures may cause nausea, vomiting, abdominal cramping, diarrhea, dehydration, and electrolyte imbalance.

Crinum asiaticum, close-up of flowers

Crinum bulbispermum, young plants showing bulbs

Management: Intravenous hydration, antiemetics, and electrolyte replacement may be necessary for patients with severe gastrointestinal symptoms, particularly in children. Consultation with a Poison Control Center should be considered. See "Poisoning by Plants with Gastrointestinal Toxins," p. 28.

References

Beutner D, Frahm AW. New alkaloids from *Crinum asiaticum*. Planta Med 1986;52:523.

Fennell CW, van Staden J. *Crinum* species in traditional and modern medicine. J Ethnopharmacol 2001;78:15–26.

Crotalaria species
Family: Leguminosae (Fabaceae)
Crotalaria fulva Roxb. (= *C. berteriana* DC.)
Crotalaria incana L.
Crotalaria juncea L.
Crotalaria lanceolata E. Mey.
Crotalaria retusa L.
Croatalaria sagittalis L.
Crotalaria spectabilis Roth

Crotalaria sagittalis

Common Names: Ala de Pico, Cascabeillo, Heliotrope, Maromera, Pete-Pete, Rabbit-Bells, **Rattle Box**, Rattleweed, Shake-Shake

Description: These coarse herbs have stout branches and sweetpea-like flowers that are usually yellow. The seeds in the persistent dried fruits cause a rattling noise when the fruit is shaken.

Distribution: *Crotalaria retusa* is cultivated in Florida. Other species of *Crotalaria* are common weeds with wide distribution in the West Indies and all parts of North America except Alaska and Canada.

Toxic Part: The whole plant is poisonous.

Toxin: Monocrotaline and other pyrrolizidine alkaloids.

Clinical Findings: Intoxication may result from contamination of grain by *Crotalaria* seeds or use of the plant in herbal teas as a folk remedy. Substantial short-term exposure (particularly to *Crotalaria fulva* and *C. juncea*) may cause acute hepatitis, and chronic exposure to lower levels may cause hepatic veno-occlusive disease (Budd–Chiari syndrome) and in some cases pulmonary hypertension.

Management: There is no known specific therapy. Consultation with a Poison Control Center should be considered. See "Poisoning by Plants with Pyrrolizidine Alkaloids," pp. 31.

References

Brooks SHE, Miller CG, McKenzie K, Avdrctsch JJ, Bras G. Acute veno-occlusive disease of the liver. Arch Pathol 1970;89:507–520.

Copple BL, Ganey PE, Roth RA. Liver inflammation during monocrotaline hepatotoxicity. Toxicology 2003;190:155–169.

Roth RA, Ganey PE. Platelets and the puzzles of pulmonary pyrrolizidine poisoning. Toxicol Appl Pharmacol 1988;93:463–471.

Stewart MJ, Steenkamp V. Pyrrolizidine poisoning: A neglected area in human toxicology. Ther Drug Monit 2001;23:698–708.

Tandon BN, Tandon RK, Tandon HD, Narndranathan M, Joshi YK. An epidemic of veno-occlusive disease of liver in central India. Lancet 1976;2:271–272.

Cryptostegia species
Family: Asclepiadaceae
Cryptostegia grandiflora R. Br.
Cryptostegia madagascariensis Bojer ex Decne.

Cryptostegia grandiflora, vine

Common Names: Purple Allamanda, Alamanda Morada Falsa, Caoutchouc, Estrella del Norte, **India Rubber Vine**, Pichuco, Rubber Vine

Description: These are woody vines that sometimes appear shrubby. The smooth, shiny green leaves are oblong and about 3 to 4 inches long. When the leaves or stems are injured, they exude a milky latex that is

Cryptostegia grandiflora, leaf showing milky latex

Cryptostegia grandiflora leaves

characteristic of this family. The flowers are purple; those of *Cryptostegia grandiflora* are more lilac, and those of *C. madagascariensis* are more red. These species may be distinguished by the length of the calyx (the green envelope holding the flower), being 0.5 inch and 1.25 inches long, respectively. Hybrids of the two species are also found. The seeds are contained in fruits typical of the Milkweed family.

Cryptostegia madagascariensis

Distribution: These plants are cultivated in south Florida and are common in the West Indies and Guam.

Toxic Part: All parts of this plant are poisonous.

Toxin: Cardioactive steroids resembling digitalis.

Clinical Findings: There are no adequately documented human poisonings, and clinical descriptions are derived primarily from animal reports. Substantial ingestion may lead to toxicity. Poisoning would be expected to produce clinical findings typical of cardioactive steroid poisoning. Toxicity has a variable latent period that depends on the quantity ingested. Dysrhythmias are usually expressed as sinus bradycardia, premature ventricular contractions, atrioventricular conduction defects, or ventricular tachydysrhythmias. Hyperkalemia, if present, may be an indicator of toxicity.

Management: Gastrointestinal decontamination as appropriate, serial electrocardiograms, and serum potassium determinations should be performed. If serious cardioactive steroid toxicity is considered, digoxin-specific Fab should be administered. Consultation with a Poison Control Center should be considered. See "Poisoning by Plants with Cardioactive Steroids/Cardiac Glycosides," p. 24.

References

Brain C, Fox VE. Suspected cardiac glycoside poisoning in elephants (*Loxodonta africana*). J S Afr Vet Assoc 1994;65:173–174.

Kamel MS, Assaf MH, Abe Y, Ohtani K, Kasai R, Yamasaki K. Cardiac glycosides from *Cryptostegia grandiflora*. Phytochemistry 2001;58:537–542.

Mathur KS, Dube BK, Kumar P. *Cryptostegia grandiflora* poisoning simulating digitalis toxicity. J Indian Med Assoc 1964;42:381–385.

Cycas species
Family: Cycadaceae
Cycas circinalis L.
Cycas revoluta Thumb.

Common Names:
Cycas circinalis: Crozier Cycas, False Sago Palm, Fern Palm, Queen Sago, **Sago Palm**
Cycas revoluta: Japanese Fern Palm, Japanese Sago Palm, **Sago Palm**

Description: *Cycas* species have stems from 3 to15 feet tall or sometimes higher (*C. revoluta* sometimes grows to 21 feet in height). They are evergreen shrubs/trees with glossy green pinnate leaves from 4 to 8 feet long. *Cycas revoluta* has basal leaflets gradually reduced to spines whereas *C. circinalis* does not have spines at the leaf base. The fleshy seeds are borne on small modified leaves.

Distribution: *Cycas circinalis* is native to India and is now widely cultivated. *Cycas revoluta*, originally native to Japan, is also widely cultivated.

Toxic Part: Seeds, leaves, unprocessed stem flour.

Cycas revoluta (above)

Cycas circinalis (left)

Toxin: Beta-methylamino-L-alanine (BMAA), a neurotoxic amino acid that is a *N*-methyl-D-aspartate (NMDA) receptor agonist (produced by a symbiotic cyanobacteria on the roots of the plants).

Clinical Findings: Acute ingestion of large amounts produces nausea, vomiting, and abdominal cramping, and diarrhea may be profound. Chronic consumption of BMAA, either directly or by eating flying foxes that have bioaccumulated the toxin by eating the seeds, is associated with a neurodegenerative disorder that is similar to amyotrophic lateral sclerosis and Alzheimer's disease.

Management: If severe gastrointestinal symptoms occur, intravenous hydration, antiemetics, and electrolyte replacement may be necessary, particularly in children. The progressive neurologic disease does not have any known therapy. Consultation with a Poison Control Center should be considered. See "Poisoning by Plants with Gastrointestinal Toxins," p. 28.

References

Chang SS, Chan YL, Wu ML, et al. Acute *Cycas* seed poisoning in Taiwan. J Toxicol Clin Toxicol 2004;42:49–54.

Cox PA, Sacks OW. Cycad neurotoxins, consumption of flying foxes, and ALS-PDC disease in Guam. Neurology 2002;58:956–959.

Murch SJ, Cox PA, Banack SA. A mechanism for slow release of biomagnified cyanobacterial neurotoxins and neurodegenerative disease in Guam. Proc Natl Acad Sci USA 2004;101:12228–12231.

Shaw CA, Wilson JM. Analysis of neurological disease in four dimensions: Insight from ALS-PDC epidemiology and animal models. Neurosci Biobehav Rev 2003;27:493–505.

Daphne mezereum L.
Family: Thymelaeaceae

Common Names: Bois Gentil, Bois Joli, **Daphne**, Dwarf Bay, February Daphne, Flax Olive, Lady Laurel, Mezereum, Mezereon, Spurge Laurel, Spurge Olive

Description: *Daphne* species are deciduous rounded shrubs that grow 4 to 5 feet in height. Leaves are elliptical, 3.5 by 0.75 inches. Flowers are lilac-purple or white and grow in clusters. They appear before the leaves. Fruits are scarlet or yellow and have a pit.

Daphne mezereum, Spring flowers and young leaves (above)

Daphne mezereum var. *album*, close-up of flowers (below)

Distribution: This Eurasian plant has escaped from cultivation in the northeastern United States and eastern Canada.

Toxic Part: The whole plant is poisonous, including the flower, but most intoxications have been associated with the fruit and seeds.

Toxin: Daphnin, the glucoside of daphnetin (a hydroxycoumarin), and mezerein (a vesicant).

Clinical Findings: There are no adequately documented human poisonings, and clinical descriptions are based on the nature of the toxin. Ingestion of the fruit or chewing the bark may produce oropharyngeal pain, vesication, and dysphagia. This may be followed by vomiting, abdominal cramping, diarrhea, dehydration, and electrolyte imbalance.

Management: Symptomatic and supportive care should be given. Consultation with a Poison Control Center is strongly suggested. See "Poisoning by Plants with Gastrointestinal Toxins," p. 28.

Daphne mezereum, fruit and flowers

Reference

Noller HG. Mezereon poisoning in a child: A contribution towards recognition of *Daphne* poisoning. Monatsschr Kinderheilkd 1955;103(7):327–330.

Datura species
Brugmansia species
Family: Solanaceae

Datura inoxia Mill. (= D. meteloides DC.)
Datura metel L. (= D. fastuosa L.)
Datura stramonium L.
Datura stramonium L. var. tatula (L.) Torrey (= D. tatula L.; this name was used for purple-flowered plants)
Datura wrightii Regel
Brugmansia arborea (L.) Lagerh.
Brugmansia × candida Pers. (= Datura × candida (Pers.) Saff.)
Datura sanguinea Ruiz & Pav. (= Brugmansia sanguinea (Ruiz & Pav.) D. Don)

Datura metel

Datura metel cv. 'Cornucopaea' (above)

Datura stramonium (below)

Datura stramonium, fruit and leaves

Brugmansia suaveolens (Humb. & Bonpl. ex Willd.) Bercht. & J. Presl. (= *Datura suaveolens* Humb. & Bonpl. ex Willd.)

Common Names:
Datura metel: **Devil's Trumpet**, Downy Thorn Apple, Horn of Plenty
Datura stramonium: Belladona de Pobre, Chamico, Concombre Zombi, Devil's Trumpet, Estramonio, Herbe aux Sorciers, Jamestown Weed, **Jimson Weed**, Kikania, La'Au-Hano, Mad Apple, Peo de Fraile, Pomme Épineuse, Stink Weed, Thorn Apple, Toloache
Datura wrightii: Indian Apple, **Sacred Thorn Apple**
Brugmansia arborea: **Angel's Trumpet**, Maikoa, Tree Datura
Brugmansia × *candida:* **Angel's Trumpet**, Belladonna, Campana, Cornucopia, Floripondio, Nana-Honua
Brugmansia suaveolens: **Florifundia**

Description: *Datura stramonium* is an annual. The stem is stout, hollow, simple, upright, or branched and grows upright to 3 to 4 feet. The leaves have a long stem, are 6 to 8 inches long, lobed, and have an offensive smell. The white flowers are funnel-form, large and showy,

Datura wrightii

Brugmansia candida, flowers

and point upward. The prickly fruits are capsules about 2 inches long, which split open along four seams to expose numerous small, kidney-shaped, brownish-black seeds when mature and dry. *Brugmansia* × *candida* may grow to 20 feet in height. The leaves are ovate, usually have smooth edges, and are pubescent. The flowers are white, yellowish-white, or pinkish-white; they are 10 to 12 inches long, funnel-form, have distinct teeth, and hang downward. The fruit of *Brugmansia* is smooth and does not split open after drying. Other species of *Datura* and *Brugmansia* generally resemble the respective foregoing descriptions.

Distribution: *Datura stramonium* is a weed throughout the West Indies, Canada, and the United States, including Hawaii but not Alaska. *Datura inoxia* grows in the West Indies, Texas, and New Mexico. *Datura metel*, a native of China, is cultivated as an ornamental and has escaped in portions of the West Indies. *Datura wrightii* is native to Mexico and the southwestern United States and is widely cultivated. *Brugmansia* species are native to South America. They are cultivated as ornamentals in the West Indies, Hawaii, and warmer regions of the United States.

Brugmansia suaveolens (above)

Datura sanguinea, flower (right)

Toxic Part: The whole plant is toxic, including the nectar; however, seeds are most often implicated in poisoning. Both the seeds and dried leaves are used to deliberately induce intoxications when a hallucinogenic action is sought. The dried leaves of *Datura stramonium* are used by herbalists in the treatment of a number of conditions, including asthma and spastic cough, even though the plant is known to be toxic.

Toxin: Atropine, scopolamine, and other anticholinergic alkaloids.

Clinical Findings: Intoxication results in dry mouth with dysphagia and dysphonia, tachycardia, and urinary retention. Elevation of body temperature may be accompanied by flushed, dry skin. Mydriasis, blurred vision, excitement and delirium, headache, and confusion may be observed.

Management: Initially, symptomatic and supportive care should be given. If the severity of the intoxication warrants intervention (hyperthermia, delirium), an antidote, physostigmine, is available. Consultation with a Poison Control Center should be considered. See "Poisoning by Plants with Anticholinergic (Antimuscarinic) Poisons," p. 21.

References

Chang SS, Wu ML, Deng JF, Lee CC, Chin TF, Liao SJ. Poisoning by *Datura* leaves used as edible wild vegetables. Vet Hum Toxicol 1999;41:242–245.

Francis PD, Clarke CF. Angel trumpet lily poisoning in five adolescents: clinical findings and management. J Paediatr Child Health 1999;35:93–95.

Isbister GK, Oakley P, Dawson AH, Whyte IM. Presumed Angel's trumpet (*Brugmansia*) poisoning: Clinical effects and epidemiology. Emerg Med (Fremantle) 2003;15:376–382.

Klein-Schwartz W, Oderda GM. Jimsonweed intoxication in adolescents and young adults. Am J Dis Child 1984;138:737–739.

Miraldi E, Masti A, Ferri S, Barni Comparini I. Distribution of hyoscyamine and scopolamine in *Datura stramonium*. Fitoterapia 2001;72:644–648.

Sopchak CA, Stork CM, Cantor RM, Ohara PE. Central anticholinergic syndrome due to Jimson weed: physostigmine therapy revisited? J Toxicol Clin Toxicol 1998;36:43–45.

Tiongson J, Salen P. Mass ingestion of Jimson Weed by eleven teenagers. Del Med J 1998;70:471–476.

Dieffenbachia species
Family: Araceae

Dieffenbachia seguine (Jacq.) Schott (Note: *D. seguine* (Jacq.) Schott 'maculata' is a cultivar [= *D. maculata* (Lodd.) G. Don; *D. picta Schott*]

Common Names: Camilichigui, Canne-á-gratter, Canne-madère, Dicha, **Dumbcane**, Dumb Plant, **Mother-in-Law's Tongue**, Pela Puerco, Rábano, Tuft Root

Description: These tall, erect, unbranched plants often are planted in pots in homes or are planted outdoors in tropical areas. Their leaves are large, oblong, pointed, and have variegated coloration; for example, different shades of green on a leaf or green splotched with ivory markings.

Distribution: Species of *Dieffenbachia* may be cultivated outdoors in southern Florida and Hawaii. These are among the most decorative pot plants for malls, offices, waiting rooms, and lobbies, as they require low light and little care.

Toxic Part: The whole plant is injurious.

Toxin: Raphides of water-insoluble calcium oxalate and unverified proteinaceous toxins.

Clinical Findings: Chewing on the leaf produces an almost immediate intense pain.

Dieffenbachia seguine

Management: The pain and edema in the oral cavity recede slowly without treatment. Cool liquids or demulcents held in the mouth may bring some relief. Analgesics may be indicated. The insoluble oxalates in these plants do not produce systemic poisoning in humans. Consultation with a Poison Control

Center should be considered. See "Poisoning by Plants with Calcium Oxalate Crystals," p. 23.

References

Chitre A, Padmanabhan S, Shastri NV. A cysteine protease of *Dieffenbachia maculata*. Indian J Biochem Biophys 1998;35:358–363.

Cumpston KL, Vogel SN, Leikin JB, Erickson TB. Acute airway compromise after brief exposure to a *Dieffenbachia* plant. J Emerg Med 2003;25:391–397.

Franceschi VR, Nakata PA. Calcium oxalate in plants: Formation and function. Annu Rev Plant Biol 2005;56:41–71.

Mrvos R, Dean BS, Krenzelok EP: *Philodendron/Dieffenbachia* ingestions: Are they a problem? Clin Toxicol 1991;29:485–491.

Pedaci L, Krenzelok EP, Jacobsen TD, Aronis J. *Dieffenbachia* species exposures: An evidence-based assessment of symptom presentation. Vet Hum Toxicol 1999;41:335–338.

Rauber A. Observations on the idioblasts of *Dieffenbachia*. J Toxicol Clin Toxicol 1985;23:79–90.

Watson JT, Jones RC, Siston AM, et al. Outbreak of food-borne illness associated with plant material containing raphides. Clin Toxicol (Phila) 2005;43:17–21.

Digitalis purpurea L.
Family: Scrophulariaceae

Common Names: Digitalis, Fairy Bells, Fairy Cap, Fairy Glove, Fairy Thimbles, Folks Glove, **Foxglove**, Ladies' Thimbles, Lion's Mouth, Pop-Dock, Rabbit Flower, Thimbles, Throatwort, Witches' Thimbles

Description: These biennial plants, growing to 3 feet tall, are commonly cultivated in gardens. Drooping flowers form along the central stalk. They may be purple, pink, or white and are usually spotted on the inside.

Digitalis purpurea

Distribution: *Digitalis* is grown in cultivation and has escaped locally. Plants are hardy in the north temperate zones including Canada and Alaska.

Toxic Part: The whole plant is poisonous. Leaves from the plant are a commercial source of the drug digitalis.

Toxin: Digitoxin, digoxin (cardioactive steroids), irritant saponins.

Digitalis purpurea, leaves

Digitalis purpurea, close-up of flowers

Clinical Findings: Pain in the mouth, nausea, emesis, abdominal pain, cramping, and diarrhea develop after ingestion. Toxicity has a variable latent period that depends on the quantity ingested. Dysrhythmias are usually expressed as sinus bradycardia, premature ventricular contractions, atrioventricular conduction defects, or ventricular tachydysrhythmias. Hyperkalemia, if present, may be an indicator of toxicity.

Management: Gastrointestinal decontamination as appropriate, serial electrocardiograms, and serum potassium determinations should be performed. If serious cardioactive steroid toxicity is considered, digoxin-specific Fab should be administered. Consultation with a Poison Control Center should be considered. See "Poisoning by Plants with Cardioactive Steroids/Cardiac Glycosides," p. 24.

References

Barrueto F, Jortani S, Valdes R, Hoffman RS, Nelson LS. Cardioactive steroid poisoning from an herbal cleansing product. Ann Emerg Med 2003;41:396–399.

Newman LS, Feinberg MW, LeWine HE. Clinical problem-solving. A bitter tale. N Engl J Med 2004;351:594–599.

Simpkiss M, Holt D. Digitalis poisoning due to the accidental ingestion of foxglove leaves. Ther Drug Monit 1983;5:217.

Slifman NR, Obermeyer WR, Aloi BK, et al. Contamination of botanical dietary supplements by *Digitalis lanata*. N Engl J Med 1998;339:806–811.

Woolf AD, Wenger T, Smith TW, Lovejoy FH Jr. The use of digoxin-specific Fab fragments for severe digitalis intoxication in children. N Engl J Med 1992;326:1739–1744.

Dirca palustris L.
Family: Thymelaeaceae

Common Names: American Mezereon, Bois de Plomb, Leather Bush, **Leatherwood**, Leaver Wood, Moosewood, Rope Bark, Swamp Wood, Wickerby Bush, Wickup, Wicopy

Description: This deciduous shrub grows to 6 feet. The leaves are 2 to 3 inches long and elliptical. Pale yellow flowers on short stalks appear from the branches before leafing. The fruit is green to reddish and about 0.25 inch in diameter.

Dirca palustris, branch with leaves (above)

Dirca palustris, close-up of flowers (below)

Distribution: *Dirca* grows in woods from New Brunswick to Ontario and Minnesota south to Florida and west to Louisiana and Oklahoma.

Toxic Part: The whole plant is toxic, particularly the bark.

Toxin: Unidentified.

Clinical Findings: No serious intoxications have been reported, probably because chewing on the plant causes immediate burning and has a nauseating taste. The bark can produce severe dermatitis, particularly during flowering and fruiting periods.

Management: Symptomatic and supportive care. Consultation with a Poison Control Center should be considered.

Reference

Ramsewak RS, Nair MG, DeWitt DL, Mattson WG, Zasada J. Phenolic glycosides from *Dirca palustris*. J Nat Prod 1999;62:1558–1561.

Duranta repens L.
(= *D. plumferi* Jacq.)
Family: Verbenaceae

Common Names: Azota Caballo, Bois Jambette, Cuentas de Oro, Garbancillo, **Golden Dewdrop**, Lila, Mais Bouilli, Pigeon Berry, Skyflower, Velo de Novia

Description: A large shrub, *Duranta* is frequently cultivated as a hedge. Flowers are small and light blue or white. The plant forms masses of persistent orange fruits.

Distribution: Golden dewdrop is native to Key West, Florida. It and related species are naturalized in southern Texas, are common in the West Indies, and are cultivated elsewhere, particularly in Hawaii and Guam.

Toxic Part: The fruit is poisonous.

Toxin: Saponin.

Clinical Findings: There are no adequately documented human poisonings, and clinical descriptions are based on the nature of the toxin. Poisoning is reputed to cause somnolence, increased body temperature, mydriasis, tachycardia, edema of lips and eyelids, and convulsions. Gastrointestinal irritation may occur.

Duranta repens, close-up of flowers and fruit (left)

Duranta repens, flowering shrub with mature fruit (below)

Management: If severe gastrointestinal symptoms occur, intravenous hydration, antiemetics, and electrolyte replacement may be necessary, particularly in children. Consultation with a Poison Control Center should be considered. See "Poisoning by Plants with Gastrointestinal Toxins," p. 28.

Reference:

Takeda Y, Morimoto Y, Matsumoto T, et al. Iridoid glucosides from the leaves and stems of *Duranta erecta*. Phytochemistry 1995;35:829–833.

Echium species
Family: Boraginaceae
Echium plantagineum L. (= E. lycopsis L.)
Echium vulgare L.

Echium vulgare

Common Names: Paterson's Curse, Blue Devil or Weed, Snake Flower, Viperine, Viper's Bugloss

Description: *Echium* are biennial plants. The erect, bristly stems grow to a height of about 2 feet and are speckled with red. The alternate leaves are oblong, prickly, and about 6 inches long. The bright blue flowers grow in recurved spikes. The fruits are small nutlets.

Distribution: These Eurasian plants are now widespread weeds in eastern North America; they are encountered infrequently west of the Mississippi but grow throughout transco tinental Canada. They are present also in Hawaii.

Toxic Part: The whole plant is poisonous.

Toxin: Pyrrolizidine alkaloids.

Clinical Findings: There are no adequately documented human poisonings, and clinical descriptions are based on the nature of the toxin. Substantial short-term exposure may cause acute hepatitis, and chronic exposure to lower levels may cause hepatic veno-occlusive disease (Budd–Chiari syndrome) and in some cases pulmonary hypertension.

Management: There is no known specific therapy. Consultation with a Poison Control Center should be considered. See "Poisoning by Plants with Pyrrolizidine Alkaloids," p. 31.

Reference

Dellow J, Seaman J. Distribution of *Echium plantagineum* and its association with pyrrolizidine alkaloid poisoning in horses in New South Wales. Plant Prot Q 1985;1:79–83.

Ephedra species
Family: Ephedraceae

Common Names: Clapweed, **Ephedra**, Joint Fir, Ma Huang, Mormon Tea

Description: This evergreen shrub ranges from 1.5 to 4 feet tall, depending on the species. It has long narrow stems and scalelike leaves with small cones borne at the nodes.

Distribution: *Ephedra sinica*, one species of *Ephedra*, is a plant native to northern China and has been utilized as a Chinese herbal preparation for

Ephedra gerardiana, close-up of stems (left)

Ephedra gerardiana (below)

thousands of years. Many other species of *Ephedra* are native to the western United States.

Toxic Part: All parts of the plant contain the active alkaloid ephedrine. However, the largest concentration is found in the stems. As a Chinese herbal remedy, ma huang is derived from the dried crushed stem of this plant.

Toxin: Ephedrine, a sympathomimetic alkaloid, as well as pseudoephedrine and norpseudoephedrine.

Clinical Findings: Ephedrine is a stimulant that is similar in chemical structure and clinical effect to amphetamine. Ingestion of this compound occurs most commonly as an herbal product and is associated with nausea, vomiting, increased heart rate, increased blood pressure, sweating, light-headedness, and weakness. Excessive dosing, such as when used for its amphetamine-like psychoactive effects, can lead to cardiovascular complications such as stroke, heart attack, heart conditions, and sudden death.

Management: Most ingestions result in little or no toxicity. Many of the amphetamine-like effects can be reduced with use of benzodiazepines, such as diazepam. In addition, symptomatic and supportive care should be instituted. The blood pressure may need to be reduced using appropriate antihypertensive agents, such as phentolamine or nitroprusside. The administration of beta-adrenergic antagonists, including labetalol, alone is contraindicated. Consultation with a Poison Control Center should be strongly considered.

References

Haller CA, Benowitz NL. Advance cardiovascular and central nervous system events associated with dietary supplements containing ephedra alkaloid. N Engl J Med 2000;343:1833–1838.

Naik SD, Freudenberger RS. Ephedra-associated cardiomyopathy. Ann Pharmacother 2004;38:400–403.

Samenuk D, Link MS, Homoud MK, et al. Adverse cardiovascular events temporally associated with ma huang, an herbal source of ephedrine. Mayo Clin Proc 2002;77:12–16.

Soni MG, Carabin IG, Griffiths JC, Burdock GA. Safety of ephedra: Lessons learned. Toxicol Lett 2004;150:97–110.

Epipremnum aureum (Linden & André) Bunt.
(= *Pothos aureus* Linden & André; *Raphidophora aurea* (Linden & André) Birdsey; *Scindapsus aureus* (Linden & André) Engl. & K. Krause)
Family: Araceae

Common Names: Amapalo, Amarillo, Devil's Ivy, Golden Ceylon Creeper, Golden Hunter's Robe, **Golden Pothos**, Hunter's Robe, Ivy Arum, Malanga, Trepadora Pothos, Solomon Island Ivy, Taro Vine, Variegated Philodendron

Description: The pothos is a climbing vine with large (up to 30 inches) heart-shaped leaves that are usually streaked with yellow.

Distribution: Pothos is an outdoor plant in the West Indies, southern Florida, Hawaii, and Guam, growing on trees for support up to a height of 40 feet. It is ubiquitous elsewhere as an indoor pot plant, either supported or in hanging baskets, where the leaves grow to only a few inches in length.

Epipremnum aureum

Toxic Part: The whole plant is injurious.

Toxin: Raphides of water-insoluble calcium oxalate and unverified proteinaceous toxins.

Clinical Findings: Chewing on the leaf produces an almost immediate intense oral pain.

Management: The pain and edema in the mouth recede slowly without treatment. Cool liquids or demulcents held in the mouth may bring some relief. Analgesics may be indicated. The insoluble oxalates in these plants do not produce systemic poisoning in man. Consultation with a Poison Control Center should be considered. See "Poisoning by Plants with Calcium Oxalate Crystals," p. 23.

References

Paulsen E, Skov P, Andersen K. Immediate skin and mucosal symptoms from pot plants and vegetables in gardeners and greenhouse workers. Contact Dermatitis 1998;39:166–170.

Rauber A. Observations on the idioblasts of *Dieffenbachia*. J Toxicol Clin Toxicol 1985;23:79–90.

Eriobotrya japonica (Thunb.) Lindl.
Family: Rosaceae

Common Names: Japanese Medlar, Japanese Plum, **Loquat**

Description: This small evergreen tree grows to about 20 feet. The large stiff leaves are rough to the touch and 8 to 12 inches in length. The fragrant, dingy

Eriobotrya japonica, young fruit

white flowers grow in clusters. The fruit is pear shaped, yellow, and about 3 inches long.

Distribution: The loquat, an edible fruit, is cultivated in California, Florida, the West Indies, the Gulf Coast states, and Hawaii.

Toxic Part: The pit kernel is toxic; the unbroken seed is harmless.

Toxin: Amygdalin, a cyanogenic glycoside.

Clinical Findings: There are no adequately documented human poisonings, and clinical descriptions are derived primarily from animal reports. Because the cyanogenic glycosides must be hydrolyzed in the gastrointestinal tract before cyanide ion is released, several hours may elapse before poisoning occurs. Abdominal pain, vomiting, lethargy, and sweating typically occur first. Cyanosis does not occur. In severe poisonings, coma develops and may be accompanied by convulsions and cardiovascular collapse.

Management: Symptomatic and supportive care should be given. Antidotal therapy is available. Consultation with a Poison Control Center is strongly suggested. See "Poisoning by Plants with Cyanogenic Compounds," p. 27.

Reference

Weber MA, Garner M. Cyanide toxicosis in Asian small-clawed otters (*Amblonyx cinereus*) secondary to ingestion of loquat (*Eriobotrya japonica*). J Zoo Wildl Med 2002;33: 145–146.

Euonymus species
Family: Celastraceae

Euonymus americanus L.
Euonymus atropurpureus Jacq.
Euonymus europaeus L.
Euonymus occidentalis Nutt.
    ex Torrey

Common Names:

Euonymus americanus: Bursting Heart, Hearts-Bursting-with-Love, **Strawberry Bush**

Euonymus atropurpureus: **Burning Bush**, Indian Arrow Wood, Wahoo

Euonymus europaeus: **European Spindle Tree**, Spindle Tree

Euonymus occidentalis: **Western Burning Bush**

Description:

Euonymus americanus: This small shrub has green stems and almost stemless leaves that are bright green and ovate, coming to a sharp point. The flowers are green-purple. The fruit is a warty capsule that is red when mature; it splits open to reveal a conspicuous, scarlet, fleshy aril in which the seed is enclosed.

Euonymus atropurpureus: This small tree branches close to the ground and has thin gray bark. The leaves are opposite and finely toothed along the

Euonymus americanus, fruit (above)

Euonymus americanus, close-up of mature fruit (below)

Euonymus atropurpureus, close-up of mature fruit

margin. Flowers and mature fruit are purple. The fruit is a four-lobed capsule with scarlet arils and a small brown seed. The fruit persists through winter.

Euonymus europaeus: The European spindle tree resembles *E. atropurpureus,* but the flowers are yellow-green, the mature capsule is pink, and the aril is bright orange.

Euonymus occidentalis: This small tree grows to 20 feet. The leaves are opposite, broadest at the base (0.5 to 2 inches), and 1.2 to 3.5 inches long; they are finely toothed on the margin. Flowers are brownish-purple. The fruit capsule has three lobes and is reddish-purple when mature. The aril is red.

Distribution:

Euonymous americanus: This species grows from southeastern New York to Florida, west to southern Illinois, Missouri, Oklahoma, and Texas.

Euonymous atropurpureus: This species grows in Ontario and New York, south to Georgia, and west to North Dakota and Texas (excluding South Carolina and Louisiana).

Euonymous europaeus: The European spindle tree was introduced from Europe. It has spread from cultivation in Massachusetts to Wisconsin and southward.

Euonymous occidentalis: The Western burning bush grows on the Pacific Coast from British Columbia to southern California.

Toxic Part: Intoxications have been reported only from the fruit of *Euonymus europaeus.* The bark from *E. atropurpureus* once was used medicinally (*Euonymus* N.F. VII).

Toxin: *Euonymus europaeus* contains a digitalis-like cardioactive steroid and a number of alkaloids with gastrointestinal irritant properties.

Clinical Findings: There are few reported human cases of human exposure, and they are not readily accessible. Ingestion of the fruit of *Euonymus europaeus* may lead to persistent emesis and watery diarrhea after a delay of 10 to 12 hours. Fever, hallucinations, somnolence, coma, and convulsions have been reported. The bark of *E. atropurpureus* is cathartic and may produce emesis.

Management: Intravenous hydration, antiemetics, and electrolyte replacement may be necessary for patients with severe gastrointestinal symptoms, particularly in children. Consultation with a Poison Control Center should be considered. See "Poisoning by Plants with Cardioactive Steroids/Cardiac Glycosides," p. 24.

Euonymus europaeus, close-up of flowers (above)

Euonymus europaeus, close-up of mature fruit (below)

References

Cooper MR, Johnson AW. *Poisonous Plants in Britain and Their Effects on Animals and Man.* Her Majesty's Stationery Office, London, England, 1984, p 305.

Frohne D, Pfander HJ. *A Colour Atlas of Poisonous Plants.* Wolfe, London, England, 1983, p 291.

Euphorbia species
Family: Euphorbiaceae
Euphorbia cotinifolia L.
Euphorbia cyathophora Murray (= *Poinsettia cyathophora* (Murray) Klotzsch &
    Garcke)
Euphorbia lactea Haw.
Euphorbia lathyris L.

Euphorbia marginata Pursh
Euphorbia milii Des Moul.
Euphorbia myrsinites L.
Euphorbia pulcherrima
Willd.ex Klotzsch
        (= *Poinsettia pulcherrima*
        (Klotzsch) Graham)
Euphorbia tirucalli L.

Common Names:
Euphorbia cotinifolia: Carrasco, Mala Mujer, **Mexican Shrubby Spurge**, Red Spurge, Yerba Lechera
Euphorbia cyathophora: Corazón de María, Fiddler's

Euphorbia cyathophora

Euphorbia lactea

Euphorbia lactea, close-up of stem

Spurge, **Fire-on-the-Mountain**, Maravilla, Mexican Fire Plant, Painted Leaf
Euphorbia lactea: **Candelabra Cactus**, Candelero, Cardón, Dragon Bones, False Cactus, Hatrack Cactus, Mottled Spurge
Euphorbia lathyris: **Caper Spurge**, Mole Plant
Euphorbia marginata: Ghostweed, Mountain Snow, **Snow-on-the-Mountain**
Euphorbia milii: Christ Plant, Christ Thorn, Corona de Cristo, Couronne du Christ, **Crown of Thorns**, Gracia de Dios
Euphorbia myrsinites: Creeping Spurge, Donkey Tail Spurge, **Myrtle Spurge**
Euphorbia pulcherrima: Christmas Flower, Christmas Star, Easter Flower, Feuilles St-Jean, Flor de Noche Buena, Flor de Pascua, Lobster Plant, Mexican Flame Leaf, Painted Leaf, **Poinsettia**
Euphorbia tirucalli: Disciplinilla, Esqueleto, Finger Tree, Indian Spurge Tree, Malabar Tree, Monkey Fiddle, Milkbush, **Pencil Tree**, Rubber Euphorbia

Euphorbia lactea (above)

Euphorbia lathyris, fruit (below)

Description: The genus *Euphorbia* contains about 1,600 species of extremely variable form; plants may appear as herbs, shrubs, or trees. Many are cactuslike with thorns. Some species contain a milky latex.

Distribution: Species of *Euphorbia* are found throughout the United States, except Alaska. There are approximately 20 species in Canada. The genus is widespread in the West Indies and Guam. Their predominant use is as house or garden plants.

Toxic Part: The latex of some species is poisonous.

Toxins: Differ based on species, but include complex diterpene esters (e.g., phorbol esters) and mitogenic lectins.

Clinical Findings: Ingestion may produce gastrointestinal symptoms. Skin exposure may produce an irritant dermatitis or be corrosive, depending on the species of *Euphorbia*. Ocular exposure may produce an irritant keratoconjunctivitis.

Most inquiries on poisoning concern the poinsettia (*Euphorbia pulcherrima*), which produces either no effect (orally or topically) or occasional cases of vomiting. This plant does not contain irritant diterpenes.

Management: Symptomatic and supportive care. Consultation with a Poison Control Center should be considered. See "Poisoning by Plants with Gastrointestinal Toxins." p. 28.

Euphorbia marginata, leaves and flowers

Euphorbia milii var. splendens, leaves and flowers

Euphorbia milii, close-up of flowers and leaves

Euphorbia pulcherrima
Euphorbia tirucalli

References

D'Archy WG. Severe contact dermatitis from poinsettia. Arch Dermatol 1974;109:909–910.

Krenzelok EP, Jacobsen TD, Aronis JM. Poinsettia exposures have good outcomes . . . just as we thought. Am J Emerg Med 1996;14:671–674.

Winek CL, Butala J, Shanor SP, Fochtman FW. Toxicology of poinsettia. Clin Toxicol 1978;13:27–45.

Ficus species
Family: Moraceae
Ficus benjamina L.
Ficus elastica Roxb. ex. Hornem.
Ficus pumila L.

Ficus benjamina, branch (above)

Ficus benjamina, close-up of leaves and fruit (below)

Common Names:
Ficus benjamina: Benjamin Tree, **Weeping Fig**, Java Fig, Laurel, Tropic Laurel, Weeping Laurel, Small-leaved Rubber Plant
Ficus elastica: Indian Rubber Tree, **Rubber Tree**, Rubber Plant, Assam Rubber
Ficus pumila: Creeping Fig, **Climbing Fig**, Creeping Rubber Plant

Description: There are about 800 species of trees, shrubs, and vines in this family, all with milky sap in their leaves and stems, some with clinging roots, native to the tropical regions, primarily of the Old World.

Ficus benjamina: This is a commonly cultivated tree or shrub, with drooping branches, and thin, leathery, ovate-elliptical leaves to 5 inches long. There are several forms, cultivated both indoors in pots and tubs and outdoors in warm areas where it can grow to 75 feet tall.
Ficus elastica: This is another commonly cultivated species, with large stems, and thick, oblong or elliptical leaves, growing to 12 inches long and glossy. The juvenile form is often cultivated indoors in pots and tubs, and outdoors this species grows to 50 feet tall when cultivated or 100 feet tall in its native habitat.
Ficus pumila: This vining species is often cultivated as a pot plant indoors or outdoors in warm areas against walls and other supporting structures. The creeping stems have leaves to 1 inch long, while on the fruiting branches the leaves are 2 to 4 inches long. The fruits are pear shaped, to 2 inches long.

Distribution: These and other *Ficus* species are commonly cultivated as pot plants in the home, tub plants in interior landscapes, and outdoors in warm climates.

Toxic Part: The sap of the plant.

Toxin: Furocoumarins, psoralens, ficin, sesquiterpenoid glucosides, and triterpines. It is unclear which of these are primarily responsible for the dermatologic effects.

Ficus elastica, branch

Clinical Findings: Exposure to the sap from the plant or leaves may produce contact dermatitis.

Management: Symptomatic and supportive care. Consultation with a Poison Control Center should be considered.

References

Kitajima J, Kimizuka K, Tanaka Y. Three new sesquiterpenoid glucosides of *Ficus pumila* fruit. Chem Pharm Bull (Tokyo) 2000;48:77–80.

Paulsen E, Skov PS, Andersen KE. Immediate skin and mucosal symptoms from pot plants and vegetables in gardeners and greenhouse workers. Contact Dermatitis 1998;39:166–170.

Ragasa CY, Juan E, Rideout JA. A triterpene from *Ficus pumila*. J Asian Nat Prod Res 1999;1:269–275.

Galanthus nivalis **L.**
Family: Amaryllidaceae

Common Names: Snowdrop

Description: The snowdrop blooms in early spring. Two grasslike leaves about 4 inches long appear with the flower. The single drooping flower is about 1 inch long and white marked with green. The fruit is a berry.

Distribution: This plant is native to Europe and is cultivated in the northern United States.

Toxic Part: The bulb is poisonous.

Toxin: Lycorine and related phenanthridine alkaloids (see *Narcissus*).

Galanthus nivalis

Clinical Findings: Ingestion of small amounts produces few or no symptoms. Large exposures may cause nausea, vomiting, abdominal cramping, diarrhea, dehydration, and electrolyte imbalance.

Management: Human poisonings are uncommon because of the small concentration of alkaloids in this plant. Intravenous hydration, antiemetics, and electrolyte replacement may be necessary for patients with severe gastrointestinal symptoms, particularly in children. Consultation with a Poison Control Center should be considered. See "Poisoning by Plants with Gastrointestinal Toxins," p. 28.

Reference

Plaitakis A, Duvoison RC. Homer's moly identified as *Galanthus nivalis* (L.): Physiologic antidote to stramonium poisoning. Clin Neuropharmacol. 1983;6:1–5.

Gelsemium sempervirens (L.) Aiton
Family: Loganiaceae

Common Names: Carolina Jasmine, Carolina Yellow Jasmine, Carolina Wild Woodbine, Evening Trumpet Flower, Madreselva, Wood Vine, **Yellow Jessamine**, Yellow False Jessamine

Description: This perennial evergreen shrubby climber has lance-shaped, paired leaves that are about 4 inches long. The flowers are funnel-form, fragrant, and bright yellow; they are occasionally solitary, but usually grow in clusters of up to 6. The fruit is a small flattened capsule with winged seeds.

Distribution: *Gelsemium* is found in woodlands from eastern Virginia to Tennessee and Arkansas, south to Florida, and west to Texas. It is also in cultivation in these regions and in southern California.

Toxic Part: Most intoxications are associated with herbal preparations of this plant; however, there are cases of children who were poisoned after sucking on the flowers.

Gelsemium sempervirens, flowering shrub (above)

Gelsemium sempervirens, close-up of flower (left)

Toxin: Gelsemine, gelsemicine, and related alkaloids.

Clinical Findings: Headache, dizziness, visual disturbances, and dry mouth with dysphagia and dysphonia result from poisoning. In severe cases, muscular weakness occurs and may be manifested by falling of the jaw and marked ptosis. There may be signs of strychnine-like action, for example, tetanic contractions and extensor spasms of the extremities following tendon taps, general rigidity, trismus, rarely convulsions.

Management: Symptomatic and supportive care should be given. Consultation with a Poison Control Center should be considered. See "Poisoning by Plants with Convulsive Poisons (Seizures)," p. 25.

Reference

Blaw ME, Adkisson MA, Levin D, Garriott JC, Tindall RS. Poisoning with Carolina jessamine (*Gelsemium sempervirens* [L.] Ait.). J Pediatr 1979;94:998–1001.

Ginkgo biloba L.
Family: Ginkgoaceae

Common Names: Ginkgo, **Maidenhair Tree**

Description: This is a very common street or ornamental deciduous tree found in the temperate zone, growing to 120 feet tall with fan-shaped green leaves 2 to 3 inches long, and yellow drupe-like fruits to 1.2 inches long that smell foul

Ginkgo biloba, close-up of branch and leaves

when ripe. Some cultivars have variegated leaves with yellow and green color.

Distribution: Native to China, it is no longer common in the wild but has been introduced to temperate regions throughout the world.

Toxic part: Seeds.

Toxin: 4-*O*-Methylpyridoxine; inhibits glutamic acid decarboxylase and prevents formation of gamma-aminobutyric acid (GABA).

Clinical Findings: Ingestion of large amounts of seeds may produce seizures, which may be delayed and may be recurrent. Typically, vomiting and diarrhea are present before the onset of seizure.

Management: If seizures occur, rapidly acting anticonvulsants, such as intravenous diazepam, should be utilized along with other supportive measures. Based on the nature of the toxin and one human case report, intravenous pyridoxine, dosed as in isoniazid poisoning, may be helpful. Consultation with a Poison Control Center should be considered. See "Poisoning by Plants with Convulsive Poisons (Seizures)," p. 25.

References

Hori Y, Fujisawa M, Shimada K, Oda A, Katsuyama S, Wada K. Rapid analysis of 4-*O*-methylpyridoxine in the serum of patients with *Ginkgo biloba* seed poisoning by ion-pair high-performance liquid chromatography. Biol Pharm Bull 2004;27:486–491.

Kajiyama Y, Fujii K, Takeuchi H, Manabe Y. Ginkgo seed poisoning. Pediatrics 2002; 109:325–327.

Miwa H, Iijima M, Tanaka S, Mizuno Y. Generalized convulsions after consuming a large amount of gingko nuts. Epilepsia 2001;42:280–281.

Sasaki K, Hatta S, Wada K, Ohshika H, Haga M. Bilobalide prevents reduction of gamma-aminobutyric acid levels and glutamic acid decarboxylase activity induced by 4-*O*-methylpyridoxine in mouse hippocampus. Life Sci 2000;67:709–715.

Gloriosa superba L.
(= *G. rothschildiana* O'Brien)
Family: Liliaceae

Common Names: Climbing Lily, **Gloriosa Lily**, Glory Lily, Pipa de Turco

Description: These climbing lilies have tuberous roots and lance-shaped leaves tipped with tendrils. The flowers are striking, bright crimson, and yellow with separated, finger-like petals. The fruits are oblong, 2–5 inches in length and contain globose red seeds.

Gloriosa superba, close-up of flowers

Distribution: The Gloriosa Lily is primarily cultivated and rarely has escaped in the extreme southern United States, the West Indies, and Hawaii.

Toxic Part: The whole plant is poisonous, particularly the tubers.

Toxin: Colchicine (*Colchicum autumnale* is a commercial source of this drug.), a cytotoxic alkaloid capable of inhibiting microtubule formation.

Clinical Findings: May cause initial oropharyngeal pain followed in several hours by intense gastrointestinal symptoms. Abdominal pain and severe, profuse, persistent diarrhea may develop, causing extensive fluid depletion and its sequelae. Colchicine may subsequently produce peripheral neuropathy, bone marrow suppression, and cardiovascular collapse.

Management: The intoxication has a prolonged course because of the slow absorption of colchicine. Aggressive symptomatic and supportive care is critical, with prolonged observation of symptomatic patients. Consultation with a Poison Control Center should be strongly considered. See "Poisoning by Plants with Mitotic Inhibitors," p. 29.

References

Aleem HMA. *Gloriosa superba* poisoning. J Assoc Physicians India 1992;40:541–542.

Mendis S. Colchicine cardiotoxicity following ingestion of *Gloriosa superba* tubers. Postgrad Med J 1989;65:752–755.

Note: There is one species for this genus, although many have been recognized in the past. (Field DV. The genus *Gloriosa*. Lilies and other Liliaceae. *Bulletin* (Royal Horticultural Society) 1973:93–95.)

Milne ST, Meek PD. Fatal colchicine overdose: Report of a case and review of the literature. Am J Emerg Med 1998;16:603–608.

Nagaratnam N, DeSilva DPKM, DeSilva N. Colchicine poisoning following ingestion of *Gloriosa superba* tubers. Trop Geogr Med 1973;25:15.

Gymnocladus dioicus (L.) K. Koch
Family: Leguminosae (Fabaceae)

Common Names: American Coffee Berry, Chicot Févier, Coffee Bean, Gros Févier, Kentucky Coffee Bean, Kentucky Mahogany, **Kentucky Coffee Tree**, Nicker Tree, Stump Tree

Description: *Gymnocladus* is a large tree that grows up to 100 feet. Leaves are alternate, twice compound, and feather like. The greenish-white flowers form in bunches. The distinctive flat, bulging fruits are 3 to 6 inches long and 1 to 2 inches wide; thick black seeds are embedded in a sticky green pulp. The pods persist through winter.

Distribution: *Gymnocladus* grows in southern Ontario south to New York and Virginia and west to eastern Nebraska; the trees are most common in western Ohio to Missouri, and are widely cultivated.

Gymnocladus dioicus, leaves

Gymnocladus dioicus, fruit with seeds

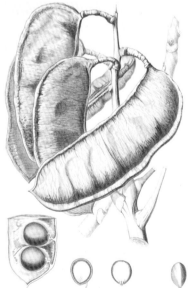

Toxic Part: Seeds and the pulp between the seeds are toxic.

Toxin: Cytisine, a nicotine-like alkaloid.

Clinical Findings: There are no adequately documented human poisonings, and clinical descriptions are based on the nature of the toxin. The cytisine content of the seeds is quite low; and chewing one or two would not be expected to produce toxic effects. The roasted seeds have been employed as a coffee substitute. Initial gastrointestinal effects may be followed by those typical of nicotine poisoning; these include hypertension, large pupils, sweating, and perhaps seizures. Severe poisoning produces coma, weakness, and paralysis that may result in death from respiratory failure.

Management: Symptomatic and supportive care should be given, with attention to adequacy of ventilation and vital signs. Atropine may reverse some of the toxic effects. Consultation with a Poison Control Center should be strongly considered. See "Poisoning by Plants with Nicotine-like Alkaloids," p. 30.

Hedera species
Family: Araliaceae
Hedera canariensis Willd.
Hedera helix L.

Common Names:
Hedera canariensis: Algerian Ivy, **Canary Ivy**, Madeira Ivy
Hedera helix: **English Ivy**, Common Ivy, Ivy, Yedra

Description: *Hedera* species are climbing vines commonly cultivated as wall covers. The leaves usually have five lobes on juvenile shoots and on root-bearing stems. Mature sections usually have nonlobed leaves on rootfree stems. Its fruits are berries that are black and globular at maturity.

Distribution: Ivy is commonly employed as an outdoor landscape plant in the northeastern United States, but is cultivated everywhere. These plants have escaped in many areas, particularly in Virginia. They are common houseplants.

Toxic Part: The berry and leaf are poisonous.

Toxin: Hederin, a saponin.

Clinical Findings: There are no adequately documented human poisonings, and clinical descriptions are based on the nature of the toxin. Most ingestions should cause little to no toxicity. The saponins are poorly absorbed, but with large exposures, gastrointestinal symptoms of nausea, vomiting, abdominal cramping, and diarrhea may occur. Allergic sensitization to this plant is common and can cause severe allergic reactions, particularly contact dermatitis.

Hedera helix, variegated form with fruit (above)

Hedera helix, in Autumn (below)

Management: If severe gastrointestinal symptoms occur, intravenous hydration, antiemetics, and electrolyte replacement may be necessary for patients with severe gastrointestinal symptoms, particularly children. Consultation with a Poison Control Center should be considered. See "Poisoning by Plants with Gastrointestinal Toxins," p. 28.

Reference

Garcia M, Fernandez E, Navarro J, del Poz MD, Fernandez de Corres L. Allergic contact dermatitis from *Hedera helix* L. Contact Dermatitis 1995;33:133–134.

Heliotropium species
Family: Boraginaceae
Heliotropium angiospermum Murray (= *H. parviflorum* L.)
Heliotropium curassavicum L.
Heliotropium indicum L.

Common Names:
Heliotropium angiospermum:
Alacrancillo, Crête de Coq, Cotorrilla, Dog's Tail, Mocos de Pavo, Rabo de Alacrán, **Scorpion's Tail**, White or Wild Clary
Heliotropium curassavicum:
Alacrancillo de Playa, Cotorrera de Playa, Kipikai, Nena, Salt Heliotrope, **Seaside Heliotrope**, Wild or Small Seaside Lavender

Heliotropium indicum, flowers

Heliotropium indicum: Alacrancillo, Cotorrera, Yerba Cotorra or Pico de Cotorra, **Indian Heliotrope**, Moco de Pavo, Scorpion Weed, Wild Clary

Description: *Heliotropium* species are fleshy herbs, mostly annuals that grow to 2 to 4 feet tall. The flowers are white, blue, purple, or pink and appear in scorpoid clusters (cymes). The fruit is a nutlet that separates into two parts, each with two nuts.

Distribution: *Heliotropium curassavicum* is widespread in the United States (except Alaska), in southwestern Canada, Mexico, the West Indies, and elsewhere throughout the world. *Heliotropium angiospermum* and *H. indicum*, an Old World plant, are pantropical species in Florida and the West Indies.

Toxic Part: The whole plant is poisonous.

Toxin: Pyrrolizidine alkaloids.

Clinical Findings: Substantial short-term exposure may cause acute hepatitis, and chronic exposure to lower quantities may cause hepatic veno-occlusive disease (Budd–Chiari syndrome) and in some cases pulmonary hypertension.

Management: There is no known specific therapy. Consultation with a Poison Control Center should be considered. See "Poisoning by Plants with Pyrrolizidine Alkaloids," p. 31.

References

Chauvin P, Dillon JC, Moren A, Talbak S, Barakaev S. Heliotrope poisoning in Tadjikistan. Lancet 1993;341:1663.

Stewart MJ, Steenkamp V. Pyrrolizidine poisoning: A neglected area in human toxicology. Ther Drug Monit 2001;23:698–708.

Tandon BN, Tandon RK, Tandon HD, et al. An epidemic of veno-occlusive disease of liver in central India. Lancet 1976;2:271–272.

Helleborus niger L.
Family: Ranunculaceae

Common Names: Christmas Rose, Black Hellebore, **Hellebore**

Description: This evergreen, early-spring-blooming perennial herb grows to 2 feet with stout rootstocks and ovate leaves that are slightly toothed at the apex. The flower is 2 to 3 inches across with five white or pinkish-white petals. The fruit is a small capsule with many glossy black seeds.

Distribution: Hellebores are of European origin and have escaped from cultivation in the northern United States and across Canada. Other *Helleborus* species, for example, *H. foetidus* L. and *H. viridis* L., also have escaped in the northeastern United States.

Helleborus niger (above)

Helleborus niger cv. 'Maximus,' flowers (below)

Toxic Part: The whole plant is poisonous.

Toxin: Hellebrin, helleborin, and helleborein, cardioactive steroids resembling digitalis.

Clinical Findings: Pain in the mouth, nausea, emesis, abdominal pain, cramping, and diarrhea develop after ingestion. Poisoning would be expected to produce Clinical findings typical of cardioactive steroid poisoning. Toxicity has a variable latent period that depends on the quantity ingested. Dysrhythmias are usually expressed as sinus bradycardia, premature ventricular contractions, atrioventricular conduction defects, or ventricular tachydysrhythmias. Hyperkalemia, if present, may be an indicator of toxicity.

Do not confuse with false hellebore (*Veratrum viride* Aiton, in Liliaceae).

Management: Gastrointestinal decontamination as appropriate, serial electrocardiograms, and serum potassium determinations should be performed. If serious cardioactive steroid toxicity is considered, digoxin-specific Fab should be administered. Consultation with a Poison Control Center should be considered. See "Poisoning by Plants with Cardioactive Steroids/Cardiac Glycosides," p. 24.

References

Bossi M, Brambilla G, Cavalli A, Marzegalli M, Regalia F. Threatening arrhythmia by uncommon digitalic toxicosis. G Ital Cardiol 1981;11:2254–2257.

Holliman A, Milton D. *Helleborus foetidus* poisoning of cattle. Vet Rec 1990;127: 339–340.

Hippobroma longiflora (L.) G. Don
(= *Isotoma longiflora* (L.) C. Presl.; *Laurentia longiflora* (L.) Peterm.)
Family: Campanulaceae

Common Names: Cipril, Feuilles Crabe, Ginbey, Horse Poison, Madame Fate, Pua-Hoku, Revienta Caballos, **Star-of-Bethlehem**, Tibey

Description: *Hippobroma* species are mostly unbranched perennial herbs with a milky latex. The simple alternate leaves are about 8 inches long. The funnel-form flower is long (2 to 3 inches), narrow (0.25 inch), and white. The fruit is a 0.75-inch-long capsule with many small brown seeds.

Distribution: This plant is a weed in the tropics, throughout the West Indies, Hawaii, and Guam.

Toxic Part: All parts of this plant are poisonous.

Toxin: Diphenyl lobelidiol, a nicotine-like alkaloid.

Hippobroma longiflora

Clinical Findings: There are no adequately documented human poisonings, and clinical descriptions are based on the nature of the toxin. Initial gastrointestinal effects may be followed by those typical of nicotine poisoning; these include hypertension, large pupils, sweating, and perhaps seizures. Severe poisoning produces coma, weakness, and paralysis that may result in death from respiratory failure.

Management: Symptomatic and supportive care should be given, with attention to adequacy of ventilation and vital signs. Atropine may reverse some of the toxic effects. Consultation with a Poison Control Center should be strongly considered. See "Poisoning by Plants with Nicotine-like Alkaloids," p. 30.

Hippomane mancinella L.
Family: Euphorbiaceae

Common Names: Manchineel Tree, Beach Apple, Mancenillier, Manzanillo

Description: This deciduous tree, native to Central America and the West Indies, grows to about 30 feet, and has thick gray bark. The leaves are elliptical and glossy, and the flowers are small, greenish-yellow to rose colored. The fruit

resembles a small crabapple and has a pleasant odor. The white latex blackens on exposure to air.

Distribution: This plant grows in the Florida Everglades and is common in the West Indies.

Toxic Part: The latex is found in all parts of the plant and is toxic.

Hippomane mancinella (above)

Hippomane mancinella, fruit (below)

Toxin: Hippomane A and B, diterpenes with a daphnane skeleton; hippomane A (or M), identical to huratoxin (from *Hura crepitans* L.).

Clinical Findings: Irritant mucosal damage upon ingestion, including severe gastrointestinal effects. The latex produces both direct and allergic contact dermatitis and keratoconjunctivitis. Inhalation of the sawdust causes cough, rhinitis, laryngitis, and bronchitis.

Management: Symptomatic and supportive care should be given. Extensive irrigation should be performed for ocular or dermal exposure. Consultation with a Poison Control Center should be considered. See "Poisoning by Plants with Gastrointestinal Toxins," p. 28.

References

Adolf W, Hecker E. On the active principles of the spurge family. X. Skin irritants, cocarcinogens, and cryptic cocarcinogens from the latex of the manchineel tree. J Nat Prod 1984;47(3):482–496.

Rao KV. Toxic principles of *Hippomane mancinella*. Planta Med 1974;25(2):166–171.

Hura crepitans L.
Family: Euphorbiaceae

Common Names: Sandbox Tree, Javillo, Molinillo, Monkey's Dinner Bell, Monkey Pistol, Possum Wood, Sablier, Salvadera

Description: This tree grows to a height of 60 feet or more with a trunk diameter of more than 3 feet at the base. The trunk and exposed roots are covered with short woody thorns. The leaves are oval, slightly serrate, and approximately 7 to 8 inches long. The green, inverted, cone-shaped male flowers are 1.5 to 2 inches long; female flowers are bright red. The woody fruits resemble a small pumpkin and are about 3 inches wide and 1.5 inches thick. The fruit pod explodes with considerable force on drying and makes a popping noise that is the basis for several of its common names. The seeds are round and flat.

Hura crepitans, flowers and leaves (above)

Hura crepitans, necklace of seed capsule parts and *Canavalia rusiosperma* seeds (below)

Distribution: *Hura crepitans* has limited distribution in southern Florida and Hawaii; it is common in the West Indies.

Toxic Part: The seeds are poisonous.

Toxin: Hurin, a plant lectin (toxalbumin) related to ricin and huratoxin, an ester of the diterpene daphnane, an irritant.

Clinical Findings: Irritant mucosal damage, including severe gastrointestinal effects, occurs on ingestion. The latex produces both direct and allergic contact dermatitis and keratoconjunctivitis. Inhalation of the sawdust causes cough, rhinitis, laryngitis, and bronchitis. Although undescribed, parenteral administration (such as by injection or inhalation), or perhaps large ingestion, may produce life-threatening systemic findings, including multisystem organ failure, even with small exposures. The patient's lymphocyte count may be elevated because of the mitogenic properties of hurin.

Management: Symptomatic and supportive care should be given. Extensive irrigation should be performed for ocular or dermal exposure. Consultation with a Poison Control Center should be strongly considered. See "Poisoning by Plants with Toxalbumins," p. 33.

References

Stirpe F, Gasperi-Campani A, Barbieri L, Falasca A, Abbondanza A, Stevens WA. Ribosome-inactivating proteins from the seeds of *Saponaria officinalis* L. (soapwort), of *Agrostemma githago* L. (corn cockle) and of *Asparagus officinalis* L. (asparagus), and from the latex of *Hura crepitans* L. (sandbox tree). Biochem J 1983;216:617–625.

Thumm EJ, Bayerl C, Goerdt S. Allergic reaction after contact with *Hura crepitans* (Sandbox tree). Hautarzt 2002;53:192–195.

Hydrangea macrophylla **(Thunb.) Ser.**
Family: Saxifragaceae

Common Names: Hydrangea, Hills-of-Snow, Hortensia, Popo-Hau, Seven Bark

Description: Hydrangea is a large bush of up to 15 feet. The stems and twigs are usually brown. The leaves are 6 inches or longer, dark green above, grayish and fuzzy beneath, and scalloped around the margin. The tiny white flowers are borne in huge clusters. The many horticultural varieties may have either rose, deep blue, or greenish-white blossoms. The flowers are persistent, turning brown as they dry. The fruit is a capsule.

Distribution: *Hydrangea* is cultivated in all parts of the United States and Canada.

Hydrangea sp.

Hydrangea macrophylla, close-up of flowers (above)

Hydrangea macrophylla, close-up of flowers (below)

Toxic Part: The flower bud is poisonous.

Toxin: Hydrangin, a cyanogenic glycoside.

Clinical Findings: There are no adequately documented human poisonings, and clinical descriptions are based on the nature of the toxin. Because the cyanogenic glycosides must be hydrolyzed in the gastrointestinal tract before cyanide ion is released, several hours may elapse before poisoning occurs. Abdominal pain, vomiting, lethargy, and sweating typically should occur first. Cyanosis does not occur. In severe poisonings, coma may develop and may be accompanied by convulsions and cardiovascular collapse. Many clinical reports describe occupational dermatitis.

Management: Symptomatic and supportive care should be given. Antidotal therapy is available. Consultation with a Poison Control Center is strongly suggested. See "Poisoning by Plants with Cyanogenic Compounds," p. 27.

Reference

Rademaker M. Occupational contact dermatitis to hydrangea. Australas J Dermatol 2003; 44:220–221.

Hydrastis canadensis L.
Family: Ranunculaceae

Common Names: Goldenseal, Orangeroot

Description: A low-growing perennial herb to 1 foot tall, with a hairy stem, and a thick yellow rhizome. The stem usually bears one 5-lobed basal leaf low to the ground and two smaller leaves near the tip, and has a small greenish-white flower. The fruits, when mature, are dark red berries.

Distribution: Found native in the deep, rich woodlands of the eastern United States and cultivated in other areas.

Toxic Part: The alkaloid can be found in all parts of the plant. The roots and rhizomes are typically ground and used as an herbal product.

Toxin: Berberine.

Clinical Findings: *Hydrastis* is most commonly ingested as a medicinal herbal product known popularly as goldenseal. Used in this form, there is a relatively low risk of acute toxicity. However, herbal usage may cause skin photosensitivity. In addition, goldenseal alters the function of cytochrome P-450 enzymes, the enzyme system involved in metabolism of common medications such as cyclosporine, digoxin, and theophylline. Berberine has cardiovascular effects

Hydrastis canadensis

Hydrastis canadensis, flower

that may be therapeutic or, at higher doses or in certain patients, toxic. Adverse cardiac effects are not yet reported in humans. When large amounts of this plant are ingested, nausea, vomiting, abdominal cramping, and diarrhea may occur.

Hydrastis canadensis, fruit

Management: Intravenous hydration, antiemetics, and electrolyte replacement may be necessary for patients with severe gastrointestinal symptoms, particularly in children. Electrocardiographic and hemodynamic monitoring seems appropriate. Consultation with a Poison Control Center should be considered. See "Poisoning by Plants with Gastrointestinal Toxins," p. 28.

References

Gurley BJ, Gardner SF, Hubbard MA, et al. In vivo effects of goldenseal, kava kava, black cohosh, and valerian on human cytochrome P450 1A2, 2D6, 2E1, and 3A4/5 phenotypes. Clin Pharmacol Ther 2005;77:415–426.

Lau CW, Yao XQ, Chen ZY, Ko WH, Huang Y. Cardiovascular actions of berberine. Cardiovasc Drug Rev 2001;19:234–244.

Palanisamy A, Haller C, Olson KR. Photosensitivity reaction in a woman using an herbal supplement containing ginseng, goldenseal, and bee pollen. J Toxicol Clin Toxicol 2003; 41:865–867.

Hymenocallis species
Family: Amaryllidaceae

Common Names: **Spider Lily**, Alligator Lily, Basket Flower, Crown Beauty, Lirio, Tararaco Blanco, Sea Daffodil

Description: The spider lilies arise from bulbs and have straplike leaves that emerge from the ground. The white or yellow flowers form in a cluster at the end of a solid, leafless stem. The fruit is a capsule.

Distribution: These plants are native to the southeastern United States and tropical America. They are popular for cultivation, primarily in tropical areas.

Toxic Part: The bulbs are poisonous.

Hymenocallis caribaea, flowers (above)

Hymenocallis declinata (below)

Toxin: Lycorine and related phenanthridine alkaloids (see *Narcissus*).

Clinical Findings: Ingestion of small amounts produces few or no symptoms. Large exposures may cause nausea, vomiting, abdominal cramping, diarrhea, dehydration, and electrolyte imbalance.

Management: Human poisonings are uncommon because of the small concentration of alkaloids in this plant. Intravenous hydration, antiemetics, and electrolyte replacement may be necessary for patients with severe gastrointestinal symptoms, particularly in children. Consultation with a Poison Control Center should be considered. See "Poisoning by Plants with Gastrointestinal Toxins," p. 28.

References

Antoun MD, Mendoza NT, Rios YR, et al. Cytotoxicity of *Hymenocallis expansa* alkaloids. J Nat Prod 1993;56:1423–1425.

Lin LZ, Hu SF, Chai HB, et al. Lycorine alkaloids from *Hymenocallis littoralis*. Phytochemistry 1995;40:1295–1298.

Hyoscyamus niger L.
Family: Solanaceae

Common Names: Henbane, Beleño, Fetid Nightshade, Insane Root, Jusquiame, Poison Tobacco, Stinking Nightshade, Black Henbane

Description: These are hairy, erect, annual or biennial weeds 2 to 6 feet high with spindle-shaped roots. The large leaves (8 inches) are coarsely and somewhat irregularly toothed. The flowers emerge singly from the main stem just above the leaves. They are greenish-yellow to yellowish-white with a purple throat and veins. The fruit is a capsule.

Distribution: Henbane was introduced from Europe and is found mostly in waste areas of the northeastern United States. It also can be found sporadically across the United States to the Pacific Coast, generally in sandy prairie. Henbane has been reported in all southern Canadian provinces. It has not been reported in Hawaii and Alaska.

Hyoscyamus niger, close-up of flower

Toxic Part: All parts of the plant are poisonous. The leaves are one source of the drug hyoscyamine.

Toxin: Atropine, scopolamine, and other anticholinergic alkaloids.

Clinical Findings: Intoxication results in dry mouth with dysphagia and dysphonia, tachycardia, and urinary retention. Elevation of body temperature may be accompanied by flushed, dry skin. Mydriasis, blurred vision, excitement and delirium, headache, and confusion may be observed.

Management: Initially, symptomatic and supportive care should be given. If the severity of the intoxication warrants intervention (hyperthermia, delirium), an antidote, physostigmine, is available. Consultation with a Poison Control Center should be considered. See "Poisoning by Plants with Anticholinergic (Antimuscarinic) Poisons," p. 21.

References

Manriquez O, Varas J, Rios JC, Concha F, Paris E. Analysis of 156 cases of plant intoxication received at the Toxicologic Information Center at Catholic University of Chile. Vet Hum Toxicol 2002;44(1):31–32.

Sands JM, Sands R. Henbane chewing. Med J Aust 1976;2:55, 58.

Spoerke DG, Hall AH, Dodson CD, Stermitz FR, Swanson CH Jr, Rumack BH. Mystery root ingestion. J Emerg Med 1987;5:385–388.

Urkin J, Shalev H, Sofer S, Witztum A. Henbane (*Hyoscyamus reticulatus*) poisoning in children in the Negev. Harefuah 1991;120:714–716.

Hypericum perforatum L.
Family: Clusiaceae

Common Name: St. John's Wort

Description: This is a perennial plant growing to 2 feet tall, with numerous upright stems, and oblong to linear leaves to 1 inch in length. The many yellow flowers are in rounded or flattened compound cymes, and have black dots near their margins. The fruit is a capsule containing many small brown seeds.

Distribution: This species is native to Europe, where it is quite common, and is naturalized throughout much of the United States and Canada, where it is found in abundance as a weed in fields, meadows, and along roads.

Hypericum perforatum

Toxic Part: Entire plant.

Toxin: Hypericin and hyperforin.

Clinical Findings: *Hypericum* is most commonly ingested as a medicinal herbal product. Used in this form, there is a relatively low risk of acute toxicity. There are reported cases of the development of the serotonin syndrome, particularly when St. John's wort is used in conjunction with other agents that increase serotonin neurotransmitter availability. Patients with the serotonin syndrome develop altered mental status, altered vital signs, muscle rigidity, and life-threatening elevation of their body temperature. Of less consequence, St. John's wort may cause skin photosensitivity. In addition, this plant may alter the function of the cytochrome P-450 enzymes, resulting in altered metabolism of certain medications.

Management: Symptomatic and supportive care. Patients with the serotonin syndrome require immediate cooling. Consultation with a Poison Control Center should be strongly considered.

Hypericum perforatum, close-up of flower

References

Brockmöller J, Reum T, Bauer S, Kerb R, Hubner WD, Roots I. Hypericin and pseudohypericin: pharmacokinetics and effects on photosensitivity in humans. Pharmacopsychiatry 1997;30:94–101.

Brown TM. Acute St. John's wort toxicity. Am J Emerg Med 2000;18:231–232.

Bryant SM, Kolodchak J. Serotonin syndrome resulting from an herbal detox cocktail. Am J Emerg Med 2004;22:625–626.

Markowitz JS, Donovan JL, DeVane CL, et al. Effect of St. John's wort on drug metabolism by induction of cytochrome P450 3A4 enzyme. JAMA 2003;290:1500–1504.

Philipp M, Kohnen R, Hiller KO. Hypericum extract versus imipramine or placebo in patients with moderate depression: Randomized multicenter study of treatment for eight weeks. BMJ 1999;319:1534–1539.

Suzuki O, Katsumata Y, Oya M. Inhibition of monoamine oxidase by hypericin. Planta Med 1984;50:272–274.

Ilex species
Family: Aquifoliaceae
Ilex aquifolium L.
Ilex opaca Aiton
Ilex vomitoria Aiton

Common Names:
Ilex aquifolium: **Holly**, English Holly, European Holly, Oregon Holly
Ilex opaca: **American Holly**
Ilex vomitoria: **Yaupon**, Appalachian or Carolina Tea, Cassena, Deer Berry, Emetic Holly, Evergreen Cassena, Indian Black Drink

Ilex opaca, flowering branch (above)

Ilex opaca, branches with fruit (below)

Ilex vomitoria, fruits

Description: *Ilex* species are evergreen trees with alternate, stiff leathery leaves. Flowers are small and white. Fruits are usually bright red at maturity.

Distribution: *Ilex opaca* is native from Massachusetts to Florida west to Missouri and Texas. *Ilex aquifolium* is cultivated from Virginia to Texas, the Pacific Coast states, and British Columbia. *Ilex vomitoria* is native to the Atlantic and Gulf Coast states and North Carolina to Texas and Arkansas.

Toxic Part: The fruit is poisonous. The leaves are nontoxic, and those of some species (e.g., *Ilex paraguariensis*, or Maté) are brewed as a beverage for their content of caffeine (or other xanthines).

Toxin: Saponins.

Clinical Findings: Most ingestions result in little or no toxicity. The saponins are poorly absorbed, but with large exposures gastrointestinal symptoms of nausea, vomiting, abdominal cramping, and diarrhea may occur. Allergic sensitization to this plant is common and can cause severe allergic reactions.

Management: If severe gastrointestinal symptoms occur, intravenous hydration, antiemetics, and electrolyte replacement may be necessary for patients with severe gastrointestinal symptoms, particularly in children. Consultation with a Poison Control Center should be considered. See "Poisoning by Plants with Gastrointestinal Toxins," p. 28.

Reference

Rodrigues TD, Johnson PN, Jeffrey LP. Holly berry ingestion: case report. Vet Hum Toxicol 1984;26:157–158.

Iris species
Family: Iridaceae
Iris germanica L.
Iris pseudacorus L.

Common Names: Flag, **Iris**, Fleur-de-Lis, Lirio Cárdeno, Orris

Description: *Iris* species are perennial herbaceous plants with a bulb or rhizome, forming two-ranked, linear to sword-shaped leaves. There are many cultivars of *Iris germanica* with showy flowers in a wide range of colors. *Iris pseudacorus* has a bright yellow flower with purple-veined leaves.

Distribution: Irises are popular garden flowers. *Iris pseudacorus*, which is native to western Europe and northern Africa, has become naturalized in eastern North America.

Toxic Part: The rootstock and leaves are toxic.

Toxin: Unidentified (may be an irritant resin).

Iris germanica, flowers

Clinical Findings: Older texts and references discuss the possibility of gastrointestinal symptoms. However, no human case reports exist substantiating this toxicity.

Management: Toxicity in humans after ingestion of this plant is unlikely. If gastrointestinal symptoms develop, intravenous hydration and antiemetics may be useful. Consultation with

Iris pseudacorus (above)

Iris pseudacorus, close-up of flowers (right)

a Poison Control Center should be considered. See "Poisoning by Plants with Gastrointestinal Toxins," p. 28.

Jatropha species
Family: Euphorbiaceae
Jatropha cathartica Terán & Berland.
Jatropha curcas L.
Jatropha gossypifolia L.
Jatropha integerrima Jacq. and varieties
Jatropha macrorhiza Benth.
Jatropha multifida L.
Jatropha podagrica Hook.

Common Names:
Jatropha cathartica: **Jicamilla**
Jatropha curcas: Barbados Nut, Cuipu, Medecinier Beni, **Physic Nut**, Piñón, Piñón Botija, Purging Nut, Ratanjyot, Tártago
Jatropha gossypifolia: **Bellyache Bush**, Frailecillo, Higuereta Cimarrona, Mala Mujer, Medicinier Barachin, Tautuba, Túatúa
Jatropha integerrima: **Peregrina**, Rose-Flowered Jatropha, Spicy Jatropha
Jatropha macrorhiza: **Jicamilla**
Jatropha multifida: **Coral Plant**, Don Tomás, Medecinier Espagnol, Papaye Sauvage, Piñón Purgante, Tártago Emético, Yuca Cimarrona

Jatropha podagrica: **Gout Stalk**, Coral Vegetal, Tinaja

Description: This is a large genus of shrubs or small trees with variously formed leaves and flowers. The flowers are usually red, and some resemble coral. Except in some Mexican species not included here, the fruit is a three-sided capsule, each section containing one seed. *Jatropha gossypifolia* usually is an annual, regardless of growing conditions; the stems of *J. macrorhiza* die aboveground in winter; all other species are perennials of tropical origin.

Jatropha curcas, branch with flowers and fruit (above)

Jatropha curcas, branch with fruit (below)

Distribution: *Jatropha cathartica* grows in Texas; *J. macrorhiza* is found in Texas, New Mexico, and Arizona. All other species are of New World origin but are now pantropical in distribution. *Jatropha* is popularly in cultivation.

Toxic Part: The seeds are poisonous. Various parts of *Jatropha curcas* are used in folk medicine.

Toxin: Jatrophin (curcin), a plant lectin (toxalbumin) related to ricin.

Clinical Findings: In contrast to poisoning with other plants containing toxic lectins, the onset of effect (nausea, vomiting, and diarrhea) is usually rapid. Other symptoms are probably secondary to fluid and electrolyte loss and suppression of intestinal function. Severe poisoning may follow ingestion of a single seed.

Jatropha podagrica, close-up of flowers and fruit (above)

Jatropha podagrica (right)

Management: Cases associated with gastrointestinal symptomatology need to be assessed for signs of dehydration and electrolyte abnormalities. Activated charcoal should be administered. Intravenous hydration, antiemetics, and electrolyte replacement may be necessary in severe cases, particularly those involving children. Consultation with a Poison Control Center should be strongly considered. See "Poisoning by Plants with Toxalbumins," p. 33.

References

Abdu-Aguye I, Sannusi A, Alafiya-Tayo RA, Bhusnurmath SR. Acute toxicity studies with *Jatropha curcas* L. Hum Toxicol 1986;5:269–274.

Gandhi VM, Cherian KM, Mulky MJ. Toxicological studies on ratanjyot oil. Food Chem Toxicol 1995;33:39–42.

Levin Y, Sherer Y, Bibi H, Schlesinger M, Hay E. Rare *Jatropha multifida* intoxication in two children. J Emerg Med 2000;19:173–175.

Kalmia species
Family: Ericaceae
Kalmia angustifolia L.
Kalmia latifolia L.
Kalmia microphylla (Hook.) A. Heller

Common Names: Mountain Laurel, American or Dwarf or Sheep or Wood Laurel, Big Leaf Ivy, Calf-, Kid-or Lamb-Kill, Calico Bush, Ivy Bush, Spoonwood, Spoonwood Ivy, Wicky

Description: These evergreen shrubs and small trees have alternate leathery leaves and white, pink, or purple flowers. The fruit is a capsule with many seeds.

Distribution: *Kalmia angustifolia* grows in eastern North America from Ontario to Labrador east to Nova Scotia and south to Michigan, Virginia, and Georgia. *Kalmia latifolia* grows in the northeastern United States south to the Gulf Coast and northern Florida to Louisana, but not in Canada. *Kalmia microphylla* occurs from Alaska to central California. These plants are found in wooded areas, moist meadows, and bogs.

Toxic Part: The leaves and nectar (in honey) are poisonous.

Toxin: Grayanotoxins (andromedotoxins), sodium channel activators.

Clinical Findings: Symptoms are predominantly neurological and cardiac. There is transient burning in the mouth after ingestion, followed after several hours by increased salivation, vomiting, diarrhea, and a tingling sensation in the skin (paresthesia). The patient may complain of headache, muscular weakness, and dimness of vision. Bradycardia and

Kalmia angustifolia (above)

Kalmia latifolia cv. 'Pink Surprise' (below)

Kalmia latifolia, fruit (below)

Kalmia latifolia, close-up of flowers

other cardiac dysrhythmias can be associated with severe blood pressure abnormalities. Coma may develop, and convulsions may be a terminal event.

Management: Fluid replacement should be instituted with respiratory support if indicated. Heart rhythm and blood pressure should be monitored and treated with appropriate medications and supportive care. Recovery is generally complete within 24 hours. Consultation with a Poison Control Center should be strongly considered. See "Poisoning by Plants with Sodium Channel Activators," p. 32.

References

Ergun K, Tufekcioglu O, Aras D, Korkmaz S, Pehlivan S. A rare cause of atrioventricular block: Mad honey intoxication. Int J Cardiol 2005;99:347–348.

Gossinger H, Hruby K, Haubenstock A, Pohl A, Davogg S. Cardiac arrhythmias in a patient with grayanotoxin-honey poisoning. Vet Hum Toxicol 1983;25:328–329.

Lampe KF. Rhododendrons, mountain laurel, and mad honey. JAMA 1988;259:2009.

Puschner B, Holstege DM, Lamberski N. Grayanotoxin poisoning in three goats. J Am Vet Med Assoc 2001;218:573–575.

Sandulescu C. A mass poisoning by honey . . . in the days of the Anabasis. Presse Med 1965;73:2070.

Yavuz H, Ozel A, Akkus I, Erkul I. Honey poisoning in Turkey. Lancet 1991;337:789–790.

Karwinskia humboldtiana (Roem. & Schult.) Zucc.
Family: Rhamnaceae

Common names: Buckthorn, Coyotillo, Tullidora

Description: This shrubby tree grows to 20 feet. Leaves are opposite and elliptical, 1 to 3 inches long and have prominent veins. The berry-shaped fruit turns black when mature and contains a pit.

Distribution: *Karwinskia* grows in western Texas, New Mexico, Arizona, and Nevada.

Toxic Part: The fruit is poisonous.

Toxin: Anthracenones (also known as buckthorn toxins: T-496, T-514, T-544) with a polycyclic polyphenol structure cause demyelination.

Clinical Findings: Weakness occurs after a latent period of several weeks, and paralysis may progress for a month or more. The syndrome may be clinically indistinguishable from Guillain–Barré syndrome. This may terminate in respiratory paralysis. Cases are reported relatively frequently in southwestern United States. Toxins from the fruit of this plant can be measured in serum by

Karwinskia humboldtiana, close-up of flowers

Karwinskia humboldtiana, branch with fruit

thin-layer chromatography (TLC) to help establish the diagnosis in unknown cases.

Management: Treatment is supportive. Mechanical ventilation may be required to treat respiratory failure caused by progressive weakness. In patients who survive, improvement progresses slowly to almost complete functional recovery. Consultation with a Poison Control Center should be strongly considered.

References

Bermudez-de Rocha MV, Lozano-Melendez FE, Tamez-Rodriguez VA, Diaz-Cvello G, Pineyro-Lopez A. The incidence of poisoning by *Karwinskia humboldtiana* in Mexico. Salud Publ Mex 1995;37:57–62.

Bermudez MV, Gonzalez-Spencer D, Guerrero M, Waksman N, Pineyro A. Experimental acute intoxication with ripe fruit *Karwinskia humboldtiana* (Tullidora) in rat, guinea pig, hamster and dog. Toxicon 1992;30:1493–1496.

Heath JW, Ueda S, Bornstein MB, Davies GD Jr, Rames CS. Buckthorn neuropathy in vitro: evidence for a primary neuronal effect. J Neuropathol Exp Neurol 1982;41:204–220.

Martinez HR, Bermudez MV, Rangel-Guerra RA, de Leon Flores L. Clinical diagnosis in *Karwinskia humboldtiana* polyneuropathy. J Neurol Sci 1998;154:49–54.

Laburnum anagyroides Medik.
(= *Cytisus laburnum* L.)
Family: Leguminosae (Fabaceae)

Common names: Golden Chain Tree, Golden Rain, Bean Tree, Laburnum

Laburnum anagyroides, close-up of flowers (right)

Laburnum anagyroides, flowering tree (below)

Description: This small tree grows to 30 feet and bears alternate, compound leaves with three leaflets and golden-yellow, sweetpea-shaped flowers in masses. Fruits contain up to eight seeds in flattened pods.

Distribution: Laburnum is widely cultivated in the southern United States.

Toxic Part: All parts of the plant, particularly the seeds, are poisonous.

Toxin: Cytisine, a nicotine-like alkaloid.

Clinical Findings: Initial gastrointestinal effects may be followed by those typical of nicotine poisoning; these include hypertension, large pupils, sweating, and perhaps seizures. Severe poisoning produces coma, weakness, and paralysis that may result in death from respiratory failure.

Management: Symptomatic and supportive care should be given, with attention to adequacy of ventilation and vital signs. Atropine may reverse some of the toxic effects. Consultation with a Poison Control Center should be strongly considered. See "Poisoning by Plants with Nicotine-like Alkaloids," p. 30.

References

Forrester RM. "Have you eaten *Laburnum*?" Lancet 1979;1:1073.

Klocking HP, Damm G, Richter M. The influence of drugs on the acute toxicity of cytisine. Arch Toxicol Suppl 1980;4:402–404.

Scane TMN, Hawkins DF. *Laburnum* "poisoning." Br Med J 1982;284:116.

Lantana camara L.
Family: Verbenaceae

Common Names: Bonbonnier or Herbe a Plomb, Cariaquillo, Cinco Negritos, Filigrana, Lakana, **Lantana**, Mikinolia-Hihiu, Shrub Verbena, Yellow Sage

Description: Lantanas are sprawling shrubs with squarish prickly stems. The opposite leaves are coarse and have a pronounced aromatic odor when crushed. Flowers grow in flat clusters. The individual flowers in a cluster change color gradually over 24 hours after opening, from yellow to orange, then to red. The outer flowers open first, and the flowers assume the appearance of concentric circles with a bull's-eye effect; the most common species has a yellow center, a surrounding circle of orange, and an outermost circle of red. Horticultural varieties may exhibit other color combinations. Fruits are small, round, and fleshy drupes, turning purple-black when mature.

Distribution: Lantanas are native to tropical America and are weeds in southern Florida, Texas, California, Hawaii, Guam, and the West Indies. In more temperate areas, these plants are cultivated as ornamentals, for example, in hanging baskets and window boxes.

Lantana camara, flowers (above)

Lantana camara, leaves and fruit (below)

Toxic Part: The immature berries are poisonous. No intoxications have been reported following ingestion of mature fruit. The leaves are also reported to be toxic to livestock.

Toxin: Unknown (probably not the phototoxin, lantadene A, as stated in some publications).

Clinical Findings: Reports of human toxicity are rare. Effects may be delayed for 2 to 6 hours after ingestion. Effects are most commonly gastrointestinal, including nausea, vomiting, abdominal cramping, and diarrhea. Severe toxicity may cause weakness, lethargy, large pupils, and respiratory depression. Reports of animal toxicity are more common. Liver toxicity (cholestasis), and photosensitization have been reported in poisoned cattle.

Management: If severe gastrointestinal effects occur, intravenous hydration, antiemetics, and electrolyte replacement may be necessary, particularly in children. Observe for abdominal pain or jaundice. Consultation with a Poison Control Center should be considered. See "Poisoning by Plants with Gastrointestinal Toxins," p. 28.

References

Morton JF. Lantana, or red sage (*Lantana camara* L., Verbenaceae), notorious weed and popular garden flower; some cases of poisoning in Florida. Econ Bot 1994;48: 259–270.

Sharma OP, Makkar HP, Dawra RK. A review of the noxious plant *Lantana camara*. Toxicon 1988;26:975–987.

Wolfson SL, Solomons TWG. Poisoning by fruit of *Lantana camara*. An acute syndrome observed in children following ingestion of the green fruit. Am J Dis Child 1964;107: 173–176.

Leucaena leucocephala (Lam.) deWit
(= *L. glauca* sense of Britton & Millsp. not (L.) Benth.)
Family: Leguminosae (Fabaceae)

Common Names: Acacia, Acacia Pálida, Aroma Blanca, Campeche, Cowbush, Ekoa, False Koa, Grains de Lin Pays, Granadino, Guacis, Hediondilla, Jimbay Bean, Jumbie Bean, Jump-and-Go, Koa-Haole, Lead Tree, **Leucaena**, Tantan, White Popinac, Wild Tamarind, Zarcilla

Description: These shrubs or small trees, growing to 30 feet tall, have feathery, twice-compound leaves with blades to 0.25 to 0.75 inch long. The flowers are creamy white or pink and grow clustered in heads that are 1 inch in diameter. Fruits are legume pods that turn reddish when mature; each pod contains 18 to 25 brown seeds.

Distribution: *Leucaena leucocephala* is native to tropical America; it grows in Florida, Texas, Hawaii, Guam, the Bahamas, and throughout the West Indies to Barbados. The seeds are commonly used for novelty jewelry.

Toxic Part: All parts of this plant are toxic, but it is commonly eaten in the West Indies and Indonesia after cooking.

Toxin: Mimosine, a heat-labile toxic amino acid, inhibits DNA synthesis through iron chelation.

Leucaena leucocephala, flower (left)

Leucaena leucocephala, branch with fruit (below)

Clinical Findings: There are no adequately documented human poisonings, and clinical descriptions are derived primarily from animal reports. The toxin is destroyed by roasting or cooking and appears to cause problems more frequently in grazing ruminants. In animals, ingestion is associated with delayed growth, esophageal lesions, and impaired thyroid function.

Management: No specific treatment is available. Poisoning is managed with general supportive care. Consultation with a Poison Control Center should be considered.

References

Hammond AC. *Leucaena* toxicosis and its control in ruminants. J Anim Sci 1995;73: 1487–1492.

Kulp KS, Vulliet PR. Mimosine blocks cell cycle progression by chelating iron in asynchronous human breast cancer cells. Toxicol Appl Pharmacol 1996;139:356–364.

Leucothoe species
Family: Ericaceae

Common Names: Dog Hobble, Dog Laurel, Fetter Bush, Pepper Bush, Switch Ivy, **Sweet Bells**, White Osier

Description: These deciduous or evergreen shrubs have simple alternate leaves and white or pink urn-shaped flowers that grow in clusters. The fruits are capsules.

Distribution: *Leucothoe* grows from Virginia to Florida, Tennessee, Louisiana, and California.

Toxic Part: The leaves and nectar (in honey) are toxic.

Toxin: Grayanotoxins (andromedotoxins), sodium channel activators.

Leucothoe sp., flowering branch

Clinical Findings: Most cases of exposure to this plant manifest minimal or no toxicity. Symptoms are predominantly neurological and cardiac. There is transient burning in the mouth after ingestion, followed after several hours by increased salivation, vomiting, diarrhea, and a tingling sensation in the skin. The patient may complain of headache, muscular weakness, and dimness of vision.

Bradycardia and other cardiac dysrhythmias can be associated with severe blood pressure abnormalities. Coma may develop, and convulsions may be a terminal event.

Management: Fluid replacement should be instituted with respiratory support if indicated. Heart rhythm and blood pressure should be monitored and treated with appropriate medications and supportive care. Recovery is generally complete within 24 hours. Consultation with a Poison Control Center should be strongly considered. See "Poisoning by Plants with Sodium Channel Activators," p. 32.

Reference

Ergun K, Tufekcioglu O, Aras D, Korkmaz S, Pehlivan S. A rare cause of atrioventricular block: Mad Honey intoxication. Int J Cardiol 2005;99:347–348.

Puschner B, Holstege DM, Lamberski N. Grayanotoxin poisoning in three goats. J Am Vet Med Assoc 2001;218:573–575.

Terai T, Araho D, Osakabe K, et al. Isolation of iso-grayanotoxin II from leaves of *Leucothoe grayana* Max. Its X-ray crystallographic analysis and acute toxicity in mice. Chem Pharm Bull 2000;48:142–144.

Yavuz H, Ozel A, Akkus I, Erkul I. Honey poisoning in Turkey. Lancet 1991;337:789–790.

Ligustrum Species
Family: Oleaceae
Ligustrum japonicum Thunb.
Ligustrum lucidum Aiton
Ligustrum vulgare L.

Ligustrum japonicum, close-up of leaves

Common Names:
Ligustrum japonicum: Wax Leaf Privet, **Japanese Privet**
Ligustrum lucidum: **Glossy Privet**, Chinese Privet, Nepal Privet, Wax Leaf Privet
Ligustrum vulgare: Hedge Plant, Lovage, Prim, **Common Privet**

Description: These semideciduous shrubs have opposite leaves, mostly oblong or ovate, with smooth margins. Small white flowers form in terminal panicles; these become large numbers of blue or black wax-coated

Ligustrum lucidum, seeds (above)

Ligustrum lucidum (right)

berries, which persist through winter. *Ligustrum japonicum* is 6 to 20 feet tall, with leathery leaves 3 to 4 inches long and flower panicles to 6 inches long. *Ligustrum lucidum* grows to 30 feet tall with shiny leves to 4 to 6 inches long and flower panicles to 10 inches long. *Ligustrum vulgare* grows to 15 feet tall with leaves to 2.5 inches long and panicles of flowers to 1.5 inches long.

Distribution: The Common Privet is native to the Mediterranean region. It is cultivated as a hedge plant throughout most of the northern United States and Canada, and it is naturalized in eastern North America. It also is cultivated in Hawaii. A number of other species

Ligustrum vulgare, close-up of flowers

having a generally similar appearance are in cultivation as well and have become naturalized, particularly *Ligustrum sinense* Lour., and are thought to be toxic.

Toxic Part: The whole plant, including the berries, is toxic.

Toxin: Syringin (ligustrin), an irritant glycoside.

Clinical Findings: Nausea, vomiting, abdominal cramping, and diarrhea may occur. There are no reported cases of poisoning in humans. In ruminants, neurotoxicity (unsteady gait, weakness) is reported to occur.

Management: If severe gastrointestinal symptoms occur, intravenous hydration, antiemetics, and electrolyte replacement may be necessary, particularly in children. Consultation with a Poison Control Center should be considered. See "Poisoning by Plants with Gastrointestinal Toxins," p. 28.

Reference

Kerr LA, Kelch WJ. Fatal privet (*Ligustrum amurense*) toxicosis in Tennessee cows. Vet Hum Toxicol 1999;41:391–392.

***Lobelia* species**
Family: Campanulaceae
Lobelia cardinalis L.
Lobelia inflata L.
Lobelia siphilitica L. (Note: Sometimes spelled *L. syphilitica*)

Common Names:
Lobelia cardinalis: Cardenal de Maceta, **Cardinal Flower**, Hog Physic, Indian Pink, Red Lobelia, Scarlet Lobelia
Lobelia inflata: Asthma Weed, Bladderpod Lobelia, Emetic Weed, Eye Bright, Gag Root, **Indian Tobacco**, Kinnikinnik, Low Belia, Puke Weed, Wild Tobacco
Lobelia siphilitica: **Blue Cardinal Flower**, Great Blue Lobelia, Great Lobelia, High Belia, Louisiana Lobelia

Lobelia cardinalis, close-up of flowers

Lobelia inflata, close-up of flowers   *Lobelia siphilitica*, close-up of flowers

Description: *Lobelia* species are erect annual or perennial herbs with alternate, simple leaves that are primarily lanceolate. The tubular flowers are distinctive in having two small petals opposed by three large petals. The flowers may be blue, pink, red, yellow, or white, in terminal racemes. The fruit is a capsule.

Distribution: The lobelias are annual weeds in most of the United States, and some (e.g., *Lobelia cardinalis*) are widely cultivated. *Lobelia inflata* is cultivated as a pharmaceutical plant. *Lobelia cardinalis* grows in damp areas (shores, meadows, and swamps) in Minnesota east to Michigan, north to Ontario, east to New Brunswick, south to eastern Texas, and east to Florida. *Lobelia inflata* grows by roadsides and in open woods in New Brunswick and Nova Scotia, west to Ontario, and south to Kansas, Arkansas, and Georgia. *Lobelia siphilitica* occurs in rich moist woods and swamps in southwest Manitoba, east of Ontario, south to Texas and Louisiana, and from Maine south to North Carolina.

Toxic Part: The whole plant is poisonous.

Toxin: Lobeline and related nicotine-like alkaloids.

Clinical Findings: Poisoning is rare and may occur when plant extracts are used as home remedies and smoking cessation aides. Initial gastrointestinal effects may be followed by those typical of nicotine poisoning; these include hypertension, large pupils, sweating, and perhaps seizures. Severe poisoning produces coma, weakness, and paralysis that may result in death from respiratory failure.

Management: Symptomatic and supportive care should be given, with attention to adequacy of ventilation and vital signs. Atropine may reverse some of the toxic effects. Consultation with a Poison Control Center should be strongly considered. See "Poisoning by Plants with Nicotine-like Alkaloids," p. 30.

References

Daly JW. Nicotinic agonists, antagonists, and modulators from natural sources. Cell Mol Neurobiol 2005;25:513–552.

Lopez R, Martinez-Burnes J, Vargas G, et al. Taxonomical, clinical and pathological findings in moradilla (*Lobelia*-like) poisoning in sheep. Vet Hum Toxicol 1994;36:195–198.

Lonicera species
Family: Caprifoliaceae
Lonicera periclymenum L.
Lonicera tatarica L.
Lonicera xylosteum L.

Common Names: Chèvrefeuille, Fly Honeysuckle, Fly Tataria, **Honeysuckle Bush**, Madreselva, Medaddy Bush, Sweet Honeysuckle, Woodbine Honeysuckle

Lonicera periclymenum, close-up of flowers (left)

Lonicera periclymenum, close-up of flowers (below)

Lonicera tatarica (above)

Lonicera tatarica, branch with fruit (below)

Description: *Lonicera* species are shrubs or climbing woody vines with opposite leaves. The flowers are tubular, two-lipped, and yellow, pink, white, or rose. The fruit is a red berry.

Distribution: Honeysuckles are extensively cultivated in the northeastern United States and Canada (Ontario, Quebec) and have escaped cultivation in these areas.

Toxic Part: Berries.

Toxin: Undescribed gastrointestinal toxins.

Clinical Findings: Older reports cite symptoms of gastrointestinal, cardiac, and neurological toxicity. Recent evidence suggests that most cases involve little or no toxicity. Substantial ingestion may cause nausea, vomiting, abdominal cramping, and diarrhea.

Management: If severe gastrointestinal symptoms occur, intravenous hydration, antiemetics, and electrolyte replacement may be necessary, particularly in children. Consultation with a Poison Control Center should be considered. See "Poisoning by Plants with Gastrointestinal Toxins," p. 28.

References

Spoerke DG, Temple AR. One year's experience with potential plant poisonings reported to the intermountain regional poison control center. Vet Hum Toxicol 1978;20:85–89.

Webster RM. Honeysuckle contact dermatitis. Cutis. 1993;51(6):424.

Lupinus species
Family: Leguminosae (Fabaceae; Papilionaceae)
Lupinus hirsutissimus Benth
Lupinus perennis L.

Common Name: Lupine

Description: This genus comprises approximately 200 species of annual and perennial herbs or small shrubs. The leaves are simple or palmately compound; the flowers are in terminal spikes or racemes, and are usually very showy in colors ranging from white to yellow, pink, blue, violet, and deep purple, and variously spotted. Fruits are a flat legume, often constricted between the seeds.

Distribution: Commonly cultivated in gardens, as well as found in the wild, particularly in western North America.

Toxic Part: The entire plant is toxic. However, ingestion of uncooked beans is the most common cause of poisoning.

Toxin: Quinolizidine alkaloids such as lupinine, lupanine, and sparteine, described variably as nicotine like or anticholinergic.

Clinical Findings: Most ingestions cause no symptoms. Ingestion of large amounts of these plants may result in dry mouth with dysphagia and dysphonia, tachycardia, and urinary retention. Elevation of body temperature may be accompanied by flushing of the skin. Mydriasis, blurred vision, excitement and delirium, headache, and confusion may be observed. Rarely, cross-allergenicity in those with peanut allergy may occur.

Management: Initially, symptomatic and supportive care should be given. If the severity of the anticholinergic syndrome warrants intervention (hyper-thermia, delirium), an antidote, physostigmine,

Lupinus perennis

is available. Consultation with a Poison Control Center should be considered. See "Poisoning by Plants with Anticholinergic (Antimuscarinic) Poisons," p. 21 and See "Poisoning by Plants with Nicotine-like Alkaloids," p. 30.

References

Di Grande A, Paradiso R, Amico S, et al. Anticholinergic toxicity associated with lupin seed ingestion: Case report. Eur J Emerg Med 2004;11:119–120.

Faeste CK, Lovik M, Wiker HG, Egaas E. A case of peanut cross-allergy to lupine flour in a hot dog bread. Int Arch Allergy Immunol 2004;135:36–39.

Smith R. Potential edible lupine poisonings in humans. Vet Hum Toxicol 1987;29:444–445.

Tsiodras S, Shin RK, Christian M, Shaw LM, Sass DA. Anticholinergic toxicity associated with lupine seeds as a home remedy for diabetes mellitus. Ann Emerg Med 1999;33:715–717.

Lycium species
Family: Solanaceae
Lycium carolinianum Walter
Lycium halmifolium Mill.

Common Names: Box Thorn, Christmas Berry, False Jessamine, **Matrimony Vine**

Description: These upright or spreading shrubs grow to 10 feet tall. The numerous branches sometimes have woody thorns. The flowers appear in

Lycium carolinianum, branch with fruit and flowers

clusters and are a dull lilac to pinkish color. The fruit is an elliptical scarlet berry about 0.75 inch in length.

Distribution: *Lycium carolinianum* is native to the southeastern United States from Texas to Florida and north to South Carolina. *Lycium halmifolium*, a Eurasian plant, is cultivated in all north temperate zones. It has escaped in the northeastern United States as far south as South Carolina and west to Tennessee and Kentucky, across southern Canada, and in California.

Toxic Part: The leaves contain the toxin.

Toxin: Damascenone, a terpenoid essential oil.

Clinical Findings: There are no adequately documented human poisonings; however, this genus belongs to a family that contains many poisonous plants. Ripe fruits are generally edible. The thorns may produce an irritant dermatitis.

Management: Symptomatic and supportive care. Consultation with a Poison Control Center should be considered.

References
Hansen AA. Stock poisoning by plants in the nightshade family. J Am Vet Med Assoc 1927;71:221.
Muenscher WC. *Poisonous Plants of the United States*, Revised Edition. Collier Books, New York, 1975.

Lycoris species
Family: Amaryllidaceae
Lycoris africana (Lam.) M. Roem. (=*L. aurea* (L'Hér.) Herb).
Lycoris radiata (L' Hér.) Herb.
Lycoris squamigera Maxim. (=*Amaryllis ballii* Hovey ex Baker)

Common Names:
Lycoris africana: Golden Hurricane Lily, **Golden Spider Lily**
Lycoris radiata: **Red Spider Lily**, Spider Lily
Lycoris squamigera: **Magic Lily**, Resurrection Lily

Description: *Lycoris* species grow from bulbs. The linear leaves emerge from the ground and die back before the flowers emerge. The red, pink, white, or yellow flowers are borne on a long, leafless stalk. The fruit is a capsule with a few smooth, black seeds.

Distribution: These plants are native to the Far East. All are cultivated extensively. *Lycoris radiata* has become naturalized in the southern United States.

Toxic Part: The bulbs are poisonous and may be mistaken for onions or scallions.

Toxin: Lycorine and related phenanthridine alkaloids (see *Narcissus*).

Clinical Findings: Ingestion of small amounts produces few or no symptoms. Large exposures may cause nausea, vomiting, abdominal cramping, diarrhea, dehydration, and electrolyte imbalance.

Management: If severe gastrointestinal effects occur, intravenous hydration, antiemetics, and electrolyte replacement may be necessary, particularly in children. Consultation with a Poison Control Center should be considered. See "Poisoning by Plants with Gastrointestinal Toxins," p. 28.

Lycoris radiata, close-up of flowers

Reference

Junko I, Akiko T, Yumiko K, et al. Poisoning by *Lycoris radiata* plants. Pharm Mon (Gekkan Yakuji) 1994;36:855–857.

Lyonia species
Family: Ericaceae

Common Names: Cereza, Clavellina, **Fetter Bush**, He Huckleberry, Male Berry, Male Blueberry, Pepper Bush, Pipe Stem, Privet Andromeda, Seedy Buckberry, Stagger Bush, Swamp Andromeda, Tetter Bush, Wicopy, Wicks

Description: These small evergreen or deciduous shrubs have alternate, simple leaves, with clusters of white or urn-shaped pink flowers. The fruit is a capsule.

Distribution: *Lyonia* grows along the Atlantic coastal plain from Rhode Island to Florida, west to Arkansas and Louisiana, and is common in the West Indies. This genus is native to Eastern Asia and North America.

Toxic Part: The leaves and nectar (in honey) are toxic.

Toxin: Lyoniol A (lyoniatoxin), chemically related to the grayanotoxins (andromedotoxins), a sodium channel activator.

Clinical Findings: There are no adequately documented human poisonings, and clinical descriptions are derived primarily from animal reports. If human toxicity were to occur, symptoms would likely resemble grayanotoxin poisoning. Symptoms are predominantly neurological and cardiac. There is transient burning in the mouth after ingestion, followed after several hours by increased

Lyonia sp., flowers

salivation, vomiting, diarrhea, and a tingling sensation in the skin. The patient may complain of headache, muscular weakness, and dimness of vision. Bradycardia and other cardiac dysrhythmias can be associated with severe blood pressure abnormalities. Coma may develop, and convulsions may be a terminal event.

Management: Fluid replacement should be instituted with respiratory support if indicated. Heart rhythm and blood pressure should be monitored and treated with appropriate medications and supportive care. Recovery is generally complete within 24 hours. Consultation with a Poison Control Center should be strongly considered. See "Poisoning by Plants with Sodium Channel Activators," p. 32.

Malus species
Family: Rosaceae

Common Names: **Apple**, Crabapple, Manzana, Pommier

Description: The apple is a deciduous tree with flowers that form in simple clusters. The fruit is a pome with seeds.

Distribution: Apple trees are widely cultivated in temperate climates. The fruit is available commercially.

Toxic Part: Seeds are poisonous.

Toxin: Amygdalin, a cyanogenic glycoside.

Malus sp., fruit

Clinical Findings: Apple seeds that are swallowed whole or chewed and eaten in small quantities are harmless. A single case of fatal cyanide poisoning has been reported in an adult who chewed and swallowed a cup of apple seeds. Because the cyanogenic glycosides must be hydrolyzed in the gastrointestinal tract before cyanide ion is released, several hours may elapse before poisoning occurs. Abdominal pain, vomiting, lethargy, and sweating typically occur first. Cyanosis does not occur. In severe poisonings, coma develops and may be accompanied by convulsions and cardiovascular collapse.

Management: Symptomatic and supportive care should be given. Antidotal therapy is available. Consultation with a Poison Control Center is strongly suggested. See "Poisoning by Plants with Cyanogenic Compounds," p. 27.

Reference

Holzbecher MD, Moss MA, Ellenberger HA. The cyanide content of laetrile preparations, apricot, peach and apple seeds. J Toxicol Clin Toxicol 1984;22:341–347.

Manihot esculenta Crantz
(=*M. utilissima* Pohl; *M. manihot* (L.) Cockerell)
Family: Euphorbiaceae

Common Names: Cassava, Juca, Manioc, Manioka, Sweet Potato Plant, Tapioca, Yuca, Yuca Brava

Description: This bushy shrub grows to 9 feet. It has milky juice and alternate leaves that are deeply parted into three to several lobes. The roots are long and tuberous.

Manihot esculenta (right)

Manihot esculenta, flowers (below)

Distribution: Cassava is cultivated in Guam, Hawaii, West Indies, Florida, and the Gulf Coast states. It is also cultivated throughout the Tropics. The so-called "sweet" variety with a lower cyanide content is most commonly found in commerce; the so-called "bitter" variety contains far more cyanide and requires extensive preparation and processing before eating.

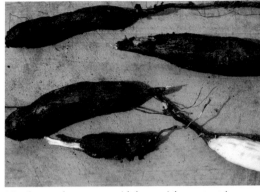

Manihot esculenta, roots with lower right root cut in half (above)

Toxic Part: Primarily the roots are poisonous; the leaves contain variable concentrations of toxin. Proper processing of the cassava root is required to remove the toxin to allow for safe consumption.

Manihot esculenta, roots in field (below)

Toxin: Linamarin and lotaustralin, cyanogenic glycosides.

Clinical Findings: Because the cyanogenic glycosides must be hydrolyzed in the gastrointestinal tract before cyanide ion is released, several hours may elapse before poisoning occurs. Abdominal pain, vomiting, lethargy, and sweating typically occur first. Cyanosis does not occur. In severe poisonings, coma develops and may be accompanied by convulsions and cardiovascular collapse.

Cyanide intake from a cassava-dominated diet has been proposed as a contributing factor to two forms of nutritional neuropathies found in Africa: tropical ataxic neuropathy and epidemic spastic paraperesis.

Management: Symptomatic and supportive care should be given. Antidotal therapy is available. Consultation with a Poison Control Center is strongly suggested. See "Poisoning by Plants with Cyanogenic Compounds," p. 27.

References

Banea-Mayambu JP, Tylleskar T, Tylleskar K, Gebre-Medhin M, Rosling H. Dietary cyanide from insufficiently processed cassava and growth retardation in children in the Democratic Republic of Congo (formerly Zaire). Ann Trop Paediatr 2000;20:34–40.

Cliff J, Lundqvist P, Martensson J, Rosling H, Sorbo B. Association of high cyanide and low sulphur intake in cassava-induced spastic paraparesis. Lancet 1985;2:1211–1213.

Espinoza OB, Perez M, Ramirez MS. Bitter cassava poisoning in eight children: A case report. Vet Hum Toxicol 1992;34:65.

Onabolu AO, Oluwole OS, Bokanga M, Rosling H. Ecological variation of intake of cassava food and dietary cyanide load in Nigerian communities. Public Health Nutr 2001;4:871–876.

Vetter J. Plant cyanogenic glycosides. Toxicon 2000;38:11–36.

Melia azedarach L.
Family: Meliaceae

Common Names: African Lilac Tree, Árbol del Quitasol, Bead Tree, **Chinaberry**, China Tree, False Sycamore, Hog Bush, Indian Lilac, Japanese Bead Tree, Inia, Lilas, Lilaila, Alelaila, Pasilla, Paradise Tree, Paraíso, Persian Lilac, Pride of China, Pride of India, Syrian Bead Tree, Texas Umbrella Tree, West Indian Lilac, White Cedar

Melia azedarach

Melia azedarach

Description: The Chinaberry is a tree that grows to 50 feet. The compound leaves are composed of many leaflets that are 2 inches long and serrated. The delicate purplish flowers are fragrant and grow in many clusters. The fruit is a 0.5-inch-wide globose berry that turns from green to yellow when mature and persists after the leaves have been shed. It contains three to five smooth, black, ellipsoidal seeds.

Distribution: The trees are native to Asia but are cultivated as ornamentals in warmer regions of the world. They have escaped widely in the south from Virginia to Florida, west to Texas, in Hawaii, and in the West Indies and Guam.

Toxic Part: The fruit and bark are poisonous. There are genetic variations in toxin content. The fruit may be eaten with impunity in some areas.

Toxin: Tetranortriterpene neurotoxins and unidentified gastrointestinal toxins.

Clinical Findings: There are no adequately documented human poisonings, and clinical descriptions are derived primarily from animal reports. There may be a prolonged latent period following ingestion. Nausea, vomiting, and diarrhea may occur as may weakness, ataxia, and tremor.

Management: If severe gastrointestinal symptoms occur, intravenous hydration, antiemetics, and electrolyte replacement may be necessary, particularly in children. Consultation with a Poison Control Center should be considered. See "Poisoning by Plants with Gastrointestinal Toxins," p. 28.

References

Hare WR. Chinaberry (*Melia azedarach*) poisoning in animals. In: Garland T, Barr AC (eds). *Toxic Plants and Other Natural Toxicants.* Cab International, Wallingford, England, 1998, pp 514–516.

Hare WR, Schutzman H, Lee BR, Knight MW. Chinaberry poisoning in two dogs. J Am Vet Med Assoc 1997;210:1638–1640.

Kiat TK. *Melia azedarach* poisoning. Singap Med J 1969;10:24–28.

Menispermum canadense L.
Family: Menispermaceae

Common Names: Canada Moonseed, **Moonseed**, Raisin de Couleuvre, Texas Sarsaparilla, Yellow Parilla, Yellow Sarsaparilla

Description: This woody twining vine grows to 12 feet. The large broad leaves are 8 inches long and slightly lobed, resembling grape leaves but with smoother edges. The grapelike fruit forms in clusters and is bluish-black with a crescent-shaped pit. The plant may be confused with wild grape.

Distribution: Moonseed is native to North America. It grows in moist wooded areas from western Quebec to Manitoba and south to Georgia and Oklahoma. Occasionally it is seen in cultivation on arbors or fences.

Menispermum canadense (above)

Menispermum canadense (below)

Toxic Part: The fruit is poisonous.

Toxin: Dauricine.

Clinical Findings: There are no adequately documented human poisonings, and clinical descriptions are based on the nature of the toxin. Dauricine possesses inhibitory effects at the cardiac potassium channel that may lead to cardiac dysrhythmia. Substantial ingestion may cause excitatory neurological symptoms including seizures, although some suggest that dauricine may produce a neuromuscular blocking effect akin to curare.

Management: Aggressive gastric decontamination. If seizures occur, rapidly acting anticonvulsants, such as intravenous diazepam, should be utilized along with other supportive measures. Dysrhythmia should be treated with standard management protocols. Consultation with a Poison Control Center should be strongly considered.

References

Doskotch R, Knapp J. Alkaloids from *Menispermum canadense*. Lloydia 1971;34:292–300.

Qian JQ. Cardiovascular pharmacological effects of bisbenzylisoquinoline alkaloid derivatives. Acta Pharmacol Sin 2002;23:1086–1092.

Xia JS, Guo DL, Zhang Y, Zhou ZN, Zeng FO, Hu CJ. Inhibitory effects of dauricine on potassium currents in guinea pig ventricular myocytes. Acta Pharmacol Sin 2000;21: 60–64.

Momordica species
Family: Cucurbitaceae
Momordica balsamina L.
Momordica charantia L.

Common Names:
Momordica balsamina: **Balsam Apple**
Momordica charantia: **Balsam Pear**, Bitter Cucumber, Bitter Gourd, Bitter Melon, Cundeamor, Fuqua, Momordique a Feuilles de Vigne, Sorci, Sorrosie, Wild Balsam Apple, Yesquin

Description: *Momordica* species are creeping vines with deeply cut leaves, trailing tendrils, and tubular yellow flowers. The warty, yellow-orange, pear-shaped or oval fruit contains bright red pulp. The fruit splits open at maturity to reveal the pulp and seeds.

Distribution: These plants are common weeds in the Gulf Coast states, Florida, the West Indies, Hawaii, and Guam.

Toxic Part: The seeds and outer rind of ripe fruit are poisonous. The red aril surrounding the seeds is edible. Boiled leaves drained of the cooking water are eaten as vegetables in the Philippines. A cultivated form of the bitter melon, much larger than the wild type, is used commonly in Asian cuisine. The leaves and fruit of *Momordica charantia* are used in folk medicine in the Caribbean and as an herbal medicine in the United States for the treatment of diabetes mellitus.

Momordica charantia, close-up of fruit, with lower right fruit opened to show mature seeds

Momordica charantia

Toxin: Momordin, a plant lectin (toxalbumin) related to ricin. Trehalose may be the antihyperglycemic principle.

Clinical Findings: The taste of *Momordica* fruit is disagreeable, and the plant lectin is relatively weak; therefore, intoxications are unlikely but potentially serious. Delayed nausea, emesis, and diarrhea result from ingestion. Hypoglycemia is unlikely but possible.

Management: Care should be taken to replace fluid and electrolyte losses. Intravenous hydration, antiemetics, glucose and electrolyte replacement may be necessary for patients with severe gastrointestinal effects, particularly in children. Consultation with a Poison Control Center should be considered. See "Poisoning by Plants with Toxalbumins," p. 33.

References

Basch E, Gabardi S, Ulbricht C. Bitter melon (*Momordica charantia*): A review of efficacy and safety. Am J Health Syst Pharm 2003;60:356–359.

Matsuur H, Asakawa C, Kurimoto M, Mizutani J. Alpha-glucosidase inhibitor from the seeds of balsam pear (*Momordica charantia*) and the fruit bodies of *Grifola frondosa*. Biosci Biotechnol Biochem 2002;66:1576–1578.

Monstera deliciosa Liebm.
Family: Araceae

Common Names: Breadfruit Vine, Casiman, Cerimán, Cerimán de Méjico, Cut Leaf Philodendron, Fruit Salad Plant, Hurricane Plant, Mexican Breadfruit, Monstera, Pinonona, Shingle Plant, **Split Leaf Philodendron**, Swiss Cheese Plant, Window Plant, Windowleaf

Description: *Monstera* species are woody stemmed climbers native to Mexico and Central America. Their large heart-shaped leaves are perforated with irregularly shaped and placed holes. A 1-foot-long, white spathe encloses a 10-inch spadix that is aromatic and edible.

Distribution: *Monstera* is native to Mexico. It is cultivated in the West Indies, Hawaii, and Guam, and grown as a greenhouse plant elsewhere.

Toxic Part: The leaves, which contain raphides of water-insoluble calcium oxalate and unverified proteinaceous toxins, are injurious. The spadix of some *Monstera* species is edible.

Clinical Findings: There are no adequately documented human poisonings, and clinical descriptions are based on the nature of the toxin. A painful burning sensation of the lips and mouth results from ingestion. There is an inflammatory reaction, often with edema and blistering. Hoarseness, dysphonia, and dysphagia may result.

Management: The pain and edema recede slowly without therapy. Cool liquids or demulcents held in the mouth may bring some relief. Analgesics may be indicated. The insoluble oxalate in these plants does not cause systemic oxalate poisoning. Consultation with a Poison Control Center should be considered. See "Poisoning by Plants with Calcium Oxalate Crystals," p. 23.

Monstera deliciosa, leaf with unripe fruit

Monstera deliciosa, mature fruit (left)

Monstera deliciosa, flower (below)

Reference

Rauber A. Observations on the idioblasts of *Dieffenbachia*. J Toxicol Clin Toxicol 1985;23: 79–90.

Myoporum laetum G. Forst.
Family: Myoporaceae

Common Names: In Australia it is called **Ngaio tree**.

Description: *Myoporum* is a large tree with a thick trunk that often has buttresses. The bark is pale grayish-white. The leaves are elliptical, coarsely toothed, and about 3 inch long. A leaf held up to the light shows numerous distinctive small dark spots (glands) that make this tree easy to identify. Many small flowers form along the terminal branches. The fruit is a small ball, less than 0.25 inch in diameter, containing a single seed.

Distribution: This tree is native to New Zealand but is planted extensively along streets in northern California.

Toxic Part: The greatest amount of toxin is contained in the leaves; however, reported intoxications in humans have involved only the fruit.

Toxin: Ngaione, an essential oil (oil of Ngaio), is a furanoid sesquiterpene ketone.

Clinical Findings: There are no adequately documented human poisonings, and clinical descriptions are derived primarily from animal reports. Substantial ingestion can produce gastrointestinal and neurological effects. Nausea, vomit-

Myoporum laetum (right)

Myoporum laetum, flower (below)

ing, abdominal cramping, and diarrhea may occur. Seizures may occur. Hepatic toxicity occurs in experimental models.

Management: Intravenous hydration, antiemetics, and electrolyte replacement may be necessary for patients with severe gastrointestinal symptoms, particularly in children. If seizures occur, rapidly acting anticonvulsants, such as intravenous diazepam, should be utilized along with other supportive measures. Consultation with a Poison Control Center should be considered. See "Poisoning by Plants with Gastrointestinal Toxins," p. 28.

References

Allen JG, Seawright AA, Hrdlicka J. The toxicity of *Myoporum tetrandrum* (Boobialla) and myoporaceous furanoid essential oils for ruminants. Aust Vet J 1978;54:287–292.

Raposo JB, Mendez MC, Riet-Correa F, de Andrade GB. Experimental intoxication by *Myoporum laetum* in sheep. Vet Hum Toxicol 1998;40:132–135.

Seawright AA, O'Donahoo RM. Light and electron microscope observations of the liver of the mouse after a single dose of ngaione. J Pathol 1972;106:251–259.

Narcissus species
Family: Amaryllidaceae
Narcissue cyclamineus DC.
Narcissus jonquilla L.
Narcissus poeticus L.
Narcissus pseudonarcissus L.
Narcissus tazetta L.

Common Names: Daffodil, Jonquil, Narciso, Narcissus, Paciencia

Description: The narcissus is native to Europe and North Africa; plants grown in the New World are mostly horticultural hybrids. The plant is grown from a bulb that is similar in structure and appearance to an onion. The leaves arise from the ground. One or more flowers arise from a stalk (scape); the blooms are usually white or yellow with

Narcissus poeticus

Narcissus pseudonarcissus L.

Narcissus pseuedonarcissus L.

six spreading petal-like parts and an erect, tubular trumpet emerging from the center.

Toxic Part: Bulb.

Toxin: Lycorine and related phenanthridine alkaloids.

Clinical Findings: Human poisoning has been associated only with ingestion of the bulbs, which were mistaken for onions. Ingestion of large quantities may produce nausea, vomiting, abdominal cramping, and diarrhea. Livestock fed narcissus bulbs in the Netherlands during wartime were also poisoned. Note that packaged bulbs sometimes contain insecticidal or antifungal agents. Contact dermatitis due to contact sensitization is a common complaint among *Narcissus* bulb handlers.

Management: Most exposures result in minimal or no toxicity. Intravenous hydration, antiemetics, and electrolyte replacement may be necessary for patients with severe gastrointestinal symptoms, particularly in children. Consultation with a Poison Control Center should be considered. See "Poisoning by Plants with Gastrointestinal Toxins," p. 28.

References

de Jong NW, Vermeulen AM, Gerth van Wijk R, de Groot H. Occupational allergy caused by flowers. Allergy 1998;53:204–209.

Gude M, Hausen BM, Heitsch H, Konig WA. An investigation of the irritant and allergenic properties of daffodils (*Narcissus pseudonarcissus* L., Amaryllidaceae). A review of daffodil dermatitis. Contact Dermatitis 1988;19:1–10.

Litovitz TL, Fahey BA. Please don't eat the daffodils. N Engl J Med 1982;306:547.

Nerium oleander L.
(=N. indicum Mill.)
Family: Apocynaceae

Common Names: Adelfa, Alhelí Extranjero, Laurier Rose, **Oleander**, 'Oliwa, 'Oleana, 'Olinana, Rosa Laurel, Rose Bay, Rosa Francesa

Description: The oleander is an evergreen shrub that grows to 20 feet and contains a clear, sticky sap. The long, leathery, narrow leaves are up to 10 inches in length and are opposite or occur in groups of three about the stem. Flowers form in small clusters at the ends of branches and are red, pink, or white. The fluffy winged seeds develop in long narrow fruits (capsules) which are 3/8 inch in diameter by 5 inches long.

Distribution: The oleander is native to the Mediterranean. It is widely cultivated outdoors in warm climates and as a tub plant elsewhere.

Toxic Part: The whole plant is toxic.

Toxin: Oleandrin, a cardioactive steroid resembling digitalis.

Clinical Findings: Poisoning has been reported from inhaling smoke from burning *Nerium*, use of the sticks to roast marshmallows, and drinking the water in which the flowers have been placed. Poisoning produces clinical findings

Nerium oleander

Nerium oleander, close-up of flower and leaves

Nerium oleander, fruits

typical of cardioactive steroid poisoning. Toxicity has a variable latent period that depends on the quantity ingested. Dysrhythmia is usually expressed as sinus bradycardia, premature ventricular contractions, atrioventricular conduction defects, or ventricular tachydysrhythmias. Hyperkalemia, if present, may be an indicator of toxicity.

Management: Gastrointestinal decontamination as appropriate, serial electrocardiograms, and serum potassium determinations should be performed. If serious cardioactive steroid toxicity is considered, digoxin-specific Fab should be administered. Consultation with a Poison Control Center should be strongly considered. See "Poisoning by Plants with Cardioactive Steroids/Cardiac Glycosides," p. 24.

References

Cheung K, Hinds JA, Diffy P. Detection of poisoning by plant-origin cardiac glycoside with the Abbott TDx analyzer. Clin Chem 1989;35:295–297.

Langford SD, Boor PJ. Oleander toxicity: An examination of human and animal toxic exposures. Toxicology 1996;109:1–13.

LeCouteur DG, Fisher AA. Chronic and criminal administration of *Nerium oleander*. J Toxicol Clin Toxicol 2002;40:523–524.

Radford DJ, Gillies AD, Hinds JA, et al. Naturally occurring cardiac glycosides. Med J Aust 1986;144:540–544.

Safadi R, Levy I, Amitai Y, Caraco Y. Beneficial effect of digoxin-specific Fab antibody fragments in oleander intoxication. Arch Intern Med 1995;55:2121–2125.

Nicotiana species
Family: Solanaceae
Nicotiana attenuata Torr. ex S. Watson
Nicotiana glauca Graham
Nicotiana longiflora Cav.
Nicotiana rustica L.
Nicotiana tabacum L.

Common Names: Paka, Tabac, Tabaco, **Tobacco**

Nicotiana glauca (above)

Nicotiana rustica (left)

Description: *Nicotiana* species may be annual or perennial; the latter generally are large shrubs or small trees. The five-lobed flowers, in large terminal panicles, are distinctively tubular, flare at the mouth, and may be white, yellow, greenish-yellow, or red. The fruit is a capsule with many minute seeds. The leaves are simple and alternate, usually have smooth edges, and often are broad, hairy, and sticky.

Nicotiana tabacum

Distribution: *Nicotiana attenuata* grows from Idaho to Baja California east to Texas. *Nicotiana glauca* is from South America and naturalized in the southwestern United States, Hawaii, Mexico, and the West Indies, and often grown as an ornamental. *Nicotiana longiflora* is commonly cultivated as a garden ornamental. It has been naturalized in the United States along the Gulf of Mexico from Texas to Florida and north to the Great Lakes.

Nicotiana rustica is from South America and has been naturalized sporadically in eastern North America from Massachusetts to Florida and in southern Ontario, and from Minnesota to New Mexico in the western United States. *Nicotiana tabacum* is the principal cultivated smoking tobacco.

Toxic Part: The whole plant is poisonous.

Toxin: The specific toxin depends on the species but involves chemically related alkaloids, for example, nicotine in *Nicotiana tabacum*, anabasine in *N. glauca*.

Clinical Findings: Acute intoxications result from ingestion of the leaves as a salad (particularly *Nicotiana glauca*), from the use of *N. tabacum* infusions in enemas as a home remedy, from the cutaneous absorption of the alkaloid during commercial tobacco harvesting, or from the ingestion of cigarettes or purified nicotine. Initial gastrointestinal symptoms may be followed by those typical of nicotine poisoning; these include hypertension, large pupils, sweating, and perhaps seizures. Severe poisoning produces coma, weakness, and paralysis that may result in death from respiratory failure.

Management: Symptomatic and supportive care should be given, with attention to adequacy of ventilation and vital signs. Atropine may reverse some of the toxic effects. Consultation with a Poison Control Center should be strongly considered. See "Poisoning by Plants with Nicotine-like Alkaloids," p. 30.

References

Arcury TA, Quandt SA, Preisser JS, Norton D. The incidence of green tobacco sickness among Latino farmworkers. J Occup Environ Med 2001;43:601–609.

Castorena JL, Garriott JC, Barnhardt FE, Shaw RF. A fatal poisoning from *Nicotiana glauca*. J Toxicol Clin Toxicol 1987;25:429–435.

Daly JW, Nicotinic agonists, antagonists, and modulators from natural sources, Cell Mol Newobial 2005;25:513–552.

Lewander WJ, Wine H, Carnevale R, et al. Ingestion of cigarettes and cigarette butts by children: Rhode Island, January 1994–July 1996. MMWR (Morb Mortal Wkly Rep) 1997; 46:125–128.

***Oenanthe aquatica* (L.) Poiret**
***Oenanthe crocata* L.**
Family: Umbelliferae (Apiaceae)

Common Names: *Oenanthe aquatica*: **Fineleaf Water Dropwort**, Water-Fennel. *Oenanthe crocata*: Dead Men's Fingers, Hemlock Water Dropwort, **Water Dropwort**

Description: *Oenanthe aquatica*: A perennial plant that has floating or prostrate stems, with multiple compound, fernlike leaves, and terminal umbels of white flowers. *Oenanthe crocata*: This perennial grows to 5 feet. Its bundle of spindle-shaped roots (dead men's fingers) contains white latex, which turns orange on exposure to air. The stem is hollow and much branched. The leaves are pinnately compound, and white flowers form in terminal umbels.

Distribution: *Oenanthe aquatica* is native to Eurasia, and the plant is known from Washington, DC, Ohio, and elsewhere. *Oenanthe crocata* is a European plant and has been introduced accidentally into marshy areas surrounding Washington, DC. *Oenanthe sarmentosa* C. Presl ex DC., which is common on the West Coast from southwest Alaska and western British Columbia to central California, is not known to be toxic.

Oenanthe crocata

Toxic Part: The whole plant is poisonous, but most intoxications have involved ingestion of the roots, which have been described as having a pleasant taste.

Toxin: Oenanthotoxin, an unsaturated aliphatic compound, similar to cicutoxin.

Clinical Findings: Onset of symptoms is rapid, usually within 1 hour of ingestion; symptoms include nausea, vomiting, salivation, and trismus. Generalized seizures also may occur. Death may occur if seizures do not terminate.

Management: Supportive care including airway management and protection against rhabdomyolysis and associated complications (e.g., electrolyte abnormalities and renal insufficiency) is the mainstay of therapy. Rapidly acting anticonvulsants (e.g., diazepam or lorazepam) for persistent seizures may be needed. Consultation with a Poison Control Center should be strongly considered. See "Poisoning by Plants with Convulsive Poisons (Seizures)," p. 25.

References

Ball MJ, Flather ML, Forfar JC. Hemlock water dropwort poisoning. Postgrad Med J 1987; 63:363–365.

Downs C, Phillips J, Ranger A, Farrell L. A hemlock water dropwort curry: A case of multiple poisoning. Emerg Med J 2002;19(5):472–473.

Mitchell MI, Routledge PA. Hemlock water dropwort poisoning: A review. J Toxicol Clin Toxicol 1978;12:417–426.

Ornithogalum species
Family: Liliaceae
Ornithogalum thyrsoides Jacq.
Ornithogalum umbellatum L.

Common Names:

Ornithogalum thyrsoides: African Wonderflower, Chincherinchee, Star-of-Beth-lehem, **Wonder Flower**

Ornithogalum umbellatum: Dove's Dung, Nap-at-Noon, **Star-of-Bethlehem**, Summer Snowflake

Description: These plants have onionlike bulbs and grasslike leaves. The flowers are borne in a cluster on an upright spike. Flowers of *Ornithogalum thyrsoides*

Ornithogalum thyrsoides

Ornithogalum thyrsoides, close-up of flowers

are white, and those of *O. umbellatum* are green and white. Seeds are contained in a capsule.

Distribution: These lilies are of Old World origin. *Ornithogalum umbellatum* has escaped and become naturalized in the southeastern United States, in the prairies of Mississippi, Missouri, Kansas, and in the northeastern United States. Both species are common garden plants.

Toxic Part: All parts of the plant are poisonous, particularly the bulb.

Ornithogalum umbellatum

Toxin: Convallatoxin and convalloside, cardioactive steroids resembling digitalis; similar to those of lily-of-the-valley (*Convallaria majalis*).

Clinical Findings: There are no adequately documented human poisonings, and clinical descriptions are derived primarily from animal reports. Substantial ingestion may lead to toxicity. Poisoning would be expected to produce clinical findings typical of cardioactive steroid poisoning. Toxicity has a variable latent period that depends on the quantity ingested. Dysrhythmias are usually expressed as sinus bradycardia, premature ventricular contractions, atrioventricular conduction defects, or ventricular tachydysrhythmias. Hyperkalemia, if present, may be an indicator of toxicity.

Management: Gastrointestinal decontamination as appropriate, serial electrocardiograms, and serum potassium determinations should be performed. If serious cardioactive steroid toxicity is considered, digoxin-specific Fab should be administered. Consultation with a Poison Control Center should be strongly considered. See "Poisoning by Plants with Cardioactive Steroids/Cardiac Glycosides," p. 24.

References

Bamhare C. Suspected cardiac glycoside intoxication in sheep and goats in Namibia due to *Ornithogalum nanodes* (Leighton). Onderstepoort J Vet Res 1988;65:25–30.

Ferth R, Kopp B. Cardenolides from *Ornithogalum umbellatum*. Pharmazie 1992;47: 626–629.

Pachyrhizus erosus (L.) Urb.
Family: Leguminosae (Fabaceae)

Common Names: Chopsui Potato, Habilla, Jícama de Agua, Jicama, Pois Cochon, Pois Manioc, Sargott, Wild Yam Bean, **Yam Bean**

Description: Yam bean is a twining vine with large tuberous roots. The violet, sweetpea-like flowers grow in racemes and produce fruits 4 to 6 inches long. Seeds are flat, square or rounded, and yellow, brown, or red.

Distribution: This plant is cultivated in the southern Gulf Coast states and has become naturalized in Florida, Hawaii, the West Indies, and Guam. The roots are commonly sold as a food in supermarkets, especially in urban areas.

Toxic Part: The seeds and mature fruit pods are toxic. The fleshy roots and the immature pods are edible.

Toxin: Gastrointestinal irritant and may contain rotenone, a mitochondrial poison.

Pachyrhizus erosus

Clinical Findings: Nausea, vomiting, abdominal cramping, and diarrhea may occur. Rotenone may produce multiple organ system failure.

Management: Intravenous hydration, antiemetics, and electrolyte replacement may be necessary for patients with severe gastrointestinal effects, particularly in children. Consultation with a Poison Control Center should be strongly considered. See "Poisoning by Plants with Gastrointestinal Toxins," p. 28.

References

De Wilde AR, Heyndrickx A, Carton D. A case of fatal rotenone poisoning in a child. J Forensic Sci 1986;31:1492–1498.

Narongchai P, Narongchai S, Thampituk S. The first fatal case of yam bean and rotenone toxicity in Thailand. J Med Assoc Thai 2005;88:984–987.

Pedilanthus tithymaloides (L.) Poit.
Family: Euphorbiaceae

Common Names: Candelilla, Christmas Candle, Devil's Backbone, Fiddle Flower, Itamo Real, Japanese Poinsettia, Jew Bush, Redbird Flower or Cactus, Ribbon Cactus, **Slipper Flower**, Slipper Plant

Description: Slipper flower is a shrubby, succulent plant characterized by numerous zigzag stems. The leaves are alternate and ovate, 2 to 4 inches long, green or with white variegations, sometimes with a pink tinge, pointed at the tip, and fall shortly after developing. Red or purple shoe-shaped "flowers" (actually floral bracts) appear on the ends of the stems. Milky white latex appears if the stem is cut. Fruits are 0.25 inches in diameter.

Distribution: This common houseplant is cultivated in warm areas and grows in the West Indies, Florida, California, Hawaii, and Guam.

Toxic Part: Milky sap found in all parts of the plants.

Pedilanthus tithymaloides (above)

Pedilanthus tithymaloides (below)

Toxin: Euphorbol and other terpenes that have gastrointestinal and dermal irritant effects.

Clinical Findings: Nausea, vomiting, abdominal cramping, and diarrhea may occur. This plant may cause very severe dermatological reactions.

Management: Intravenous hydration, antiemetics, and electrolyte replacement may be necessary for patients with severe gastrointestinal symptoms, particularly in children. Consultation with a Poison Control Center should be considered. See "Poisoning by Plants with Gastrointestinal Toxins," p. 28.

Reference

Lim TK, Soepadmo E. Eye injury from plant sap of *Pedilanthus tithymaloides* Poit. Singap Med J 1984;25:412–419.

Pentalinon luteum (L.) B.F. Hansen & R.P. Wunderlin.
(=*Urechites lutea* (L.) Britton)
Family: Apocynaceae

Common Names: Babeiro Amarillo, Bejuco Ahoja Vaca, Catesby's Vine, Corne Cabrite, Curamaguey, Nightsage, Wild Allamanda, Wild Nightshade, Wild Unction, **Yellow Nightshade**

Description: Yellow nightshade is a shrubby vine with milky sap. The opposite leaves are 2 inches wide and 3.5 inches long, obtuse at both the tip and base. The flowers grow in a cluster (cyme), and only a few are open at one time. There are five yellow petals, which may be marked with red on the inside. The fruit usually appears in pairs as thin, woody, slightly curved pods up to 8 inches long. The pod contains winged seeds.

Distribution: This plant grows in Florida, the Bahamas, and the Greater and Lesser Antilles south to St. Vincent.

Pentalinon luteum

Toxic Part: The leaf is poisonous.

Toxin: Urechitoxin, a cardioactive steroid resembling digitalis.

Clinical Findings: There are no adequately documented human poisonings, and clinical descriptions are derived primarily from animal reports. Substantial ingestion may lead to

Pentalinon luteum

toxicity. Poisoning would be expected to produce clinical findings typical of cardioactive steroid poisoning. Toxicity has a variable latent period that depends on the quantity ingested. Symptoms are usually expressed as sinus bradycardia, premature ventricular contractions, atrioventricular conduction defects, or ventricular tachydysrhythmias. Hyperkalemia, if present, may be an indicator of toxicity.

Management: Gastrointestinal decontamination as appropriate, serial electrocardiograms, and serum potassium determinations should be performed. If serious cardioactive steroid toxicity is considered, digoxin-specific Fab should be administered. Consultation with a Poison Control Center should be considered. See "Poisoning by Plants with Cardioactive Steroids/Cardiac Glycosides," p. 24.

References

Faz EM. *Urechites lutea* (L.) Britton toxicity in cattle. Vet Hum Toxicol 1996;38: 313–314.

Marrero E, Alfonso HA, Fajardo M. Cardiotonic effect of different glycoside extracts of *Urechites lutea*. Rev Salud Anim 1985;7:79–90.

Pernettya species
Family: Ericaceae

Common Name: Pernettya.

Description: *Pernettya* species are small evergreen shrubs with simple, alternate, leathery leaves. The white urn-shaped flowers usually appear singly. The fruit is a white or brightly colored berry that persists through the winter.

Distribution: *Pernettya* species are native to Central and South America, New Zealand, and Tasmania. They are only in cultivation in the United States.

Pernettya mucronata, close-up of flowers

Pernettya mucronata, branch with fruit

Toxic Part: Leaves, berries, nectar from flowers (in honey).

Toxin: Grayanotoxins (andromedotoxins), sodium channel activators.

Clinical Findings: There are no adequately documented human poisonings, and clinical descriptions are based on the nature of the toxin. Symptoms are expected to be predominantly neurological and cardiac. There may be transient burning in the mouth after ingestion, followed after several hours by increased salivation, vomiting, diarrhea, and a tingling sensation in the skin. The patient may complain of headache, muscular weakness, and dimness of vision. Bradycardia and other cardiac dysrhythmias can be associated with severe blood pressure abnormalities. Coma may develop, and convulsions may be a terminal event.

Management: Fluid replacement should be instituted with respiratory support if indicated. Heart rhythm and blood pressure should be monitored and treated with appropriate medications and supportive care. Recovery is generally complete within 24 hours. Consultation with a Poison Control Center should be strongly considered. See "Poisoning by Plants with Sodium Channel Activators," p. 32.

References

Ergun K, Tufekcioglu O, Aras D, Korkmaz S, Pehlivan S. A rare cause of atrioventricular block: Mad Honey intoxication. Int J Cardiol 2005;99:347–348.

Masutani T, Seyama I, Narahashi T, Iwasa J. Structure-activity relationship for grayanotoxin derivatives in frog skeletal muscle. J Pharmacol Exp Ther 1981;217:812–819.

Puschner B, Holstege DM, Lamberski N. Grayanotoxin poisoning in three goats. J Am Vet Med Assoc 2001;218:573–575.

Yavuz H, Ozel A, Akkus I, Erkul I. Honey poisoning in Turkey. Lancet 1991;337:789–790.

Philodendron species
Family: Araceae

Common Names: Bejuco de Lombriz, Paisaje, **Philodendron**

Description: Philodendrons are mostly climbing vines with aerial roots. The adult leaves are often large and variable, and two types are encountered: the first resembles *Alocasia* but has more pronounced irregular notching around the leaf margin, and the other form (e.g., *Philodendron scandens* C. Koch & H. Sello and its variants), which is widely available, has heart-shaped or oblong leaves with smooth margins, sometimes with variegated patterns. Juvenile leaves are smaller and do not vary much by species, making identification at this stage difficult.

Distribution: Philodendron grows outdoors in warm climates. It is probably the most popular indoor houseplant in the United States.

Toxic Part: The leaves are injurious.

Toxin: Insoluble calcium oxalate crystals.

Philodendron sp., flowers (left)

Philodendron selloum (below)

Clinical Findings: Although this plant contains calcium oxalates, they are not well organized and therefore not typically associated with the same toxicity as dumbcane (*Dieffenbachia* species). Substantial ingestion may cause some gastrointestinal irritation including nausea, vomiting, and diarrhea. Most cases result in no symptoms.

Management: Most cases result in no toxicity. Intravenous hydration, antiemetics, and electrolyte replacement may be necessary for patients with severe gastrointestinal symptoms, particularly in children. Consultation with a Poison Control Center should be considered. See "Poisoning by Plants with Calcium Oxalate Crystals," p. 23.

References

Franceschi VR, Nakata PA. Calcium oxalate in plants: Formation and function. Annu Rev Plant Biol 2005;56:41–71.

McIntire MS, Guest JR, Porterfield JF. *Philodendron*—an infant death. J Toxicol Clin Toxicol 1990;28:177–183.

Mrvos R, Dean BS, Krenzelok EP. *Philodendron/Dieffenbachia* ingestions: Are they a problem? J Toxicol Clin Toxicol 1991;29:485–491.

Phoradendron **species**
Family: Loranthaceae
Phoradendron quadrangulare (Kunth) Griseb. (Note: *P. rubrum* misapplied)
Phoradendron serotinum (Raf.) M.C. Johnst. var. *serotinum* (Note: *P. flavescens* misapplied)
Phoradendron serotinum (Raf.) M.C. Johnst. var. *tomentosum* (DC.) Kuijt

Common Names:
Phoradendron quadrangulare: **Cepa Caballero**
Phoradendron serotinum: **American Mistletoe**, False Mistletoe, **Mistletoe**
Phoradendron tomentosum: **Injerto**

Description: These species are parasitic plants that grow on the trunks and branches of trees. They have thick, leathery, smooth-edged, opposite leaves and tiny flowers that grow in small spikes. The fruits are globose white berries and translucent in *Phoradendron serotinum* and *P. tomentosum*, and pinkish in *P. quadrangulare*.

Distribution: *Phoradendron quadrangulare* is parasitic only on the mahogany (*Swietenia mahogani*) throughout its entire range: southernmost Florida, the Bahamas, Cuba, Jamaica, Haiti, Dominican Republic, and Puerto Rico. *Phoradendron serotinum* is the mistletoe plant sold at Christmas. It is parasitic on deciduous trees in the southeast from New Jersey to Florida, west to southern Illinois and Texas. *Phoradendron tomentosum* grows from Kansas to

Louisiana, west to Texas and into Mexico.

Toxic Part: The leaves and stems are toxic. The berries may be toxic if consumed in large amounts.

Toxin: Phoratoxin, a plant lectin (toxalbumin) related to ricin.

Clinical Findings: Most exposures result in minimal or no toxicity. However, the presence of phoratoxin raises concerns following the ingestion of large amounts. Initial symptoms depend on the amount of toxin exposure and include nausea, vomiting, abdominal cramping, diarrhea, and dehydration. More severe toxicity involves multiple organ system failures.

Management: Cases associated with gastrointestinal efforts need to be assessed for signs of dehydration and electrolyte abnormalities. Activated charcoal should be administered. Intravenous hydration, antiemetics, and electrolyte replacement may be necessary in severe cases,

Phoradendron serotinum, growing on tree (above)

Phoradendron serotinum (below)

particularly those involving children. Consultation with a Poison Control Center should be strongly considered. See "Poisoning by Plants with Toxalbumins," p. 33.

References

Hall AH, Spoerke DG, Rumack BH. Assessing mistletoe toxicity. Ann Emerg Med 1986;15: 1320–1323.

Krenzelok EP, Jacobsen TD, Aronis J. American mistletoe exposures. Am J Emerg Med 1997;15:516–520.

Spiller HA, Willias DB, Gorman SE, Sanftleban J. Retrospective study of mistletoe ingestion. J Toxicol Clin Toxicol 1996;34:405–408.

Physalis species
Family: Solanaceae

Common Names: Alquerquenje, Barbados Gooseberry, Battre Autour, Cape Gooseberry, Chinese Lantern Plant, Coque Molle, Coqueret, Farolito, Ground Cherry, Gooseberry Tomato, Huevo de Gato, Husk Tomato, Jamberry, Maman Laman, **Japanese Lantern Plant**, Mexican Husk Tomato, Pa'Ina, Poha, Sacabuche, Strawberry Tomato, Tomates, Tope-Tope, Vejiga de Perro, Winter Cherry, Yellow Henbane

Description: There are about 17 species of *Physalis* growing in the United States. A large number are cultivated for their attractive Chinese lantern-like fruit pods. Inside the paper-like "pod" (enlarged calyx) is a globose fruit (berry) with minute seeds. When mature, the raw or cooked berries of some species are edible.

Physalis alkekengi

Distribution: *Physalis* plants are both native and cultivated throughout the United States (including Hawaii), central and eastern Canada, the West Indies, and Guam.

Toxic Part: The unripe berries are poisonous.

Toxin: Solanine glycoalkaloids have predominantly gastrointestinal irritant effects.

Clinical Findings: Nausea, vomiting, abdominal cramping, and diarrhea may occur. Central nervous system effects of delerium, hallucinations, and coma have been reported, but the mechanism for these effects is not known.

Physalis alkekengi, fruit cut in half (above)

Physalis crassifolia (below)

Management: Intravenous hydration, antiemetics, and electrolyte replacement may be necessary for patients with severe gastrointestinal effects, particularly in children. Central nervous system effects are managed with supportive and symptomatic care. Consultation with a Poison Control Center should be considered. See "Poisoning by Plants with Gastrointestinal Toxins," pp. 28.

References

McMillan M, Tompson JC. An outbreak of suspected solanine poisoning in schoolboys. Examination of solanine poisoning. Q J Med 1979;48:227–243.

Slanina P. Solanine (glycoalkaloids) in potatoes: Toxicological evaluation. Food Chem Toxicol 1990;28:759–761.

Phytolacca americana L. (=*P. decandra* L.)
Family: Phytolaccaceae

Phytolacca americana, branch with fruit

Common Names: American Nightshade, Bledo Carbonero, Cancer Jalap, Chongrass, Cocum, Cokan, Coakum, Crow Berry, Garget, Indian Polk, Ink Berry, Pigeonberry, Pocan Bush, Poke, Pokeberry, **Pokeweed**, Red Ink Plant, Red Weed, Scoke

Description: The pokeweed has a large perennial rootstock (up to 6 inches in diameter) from which stout, purplish, branching stems emerge up to 12 feet in height. The plant has a strong, unpleasant odor. The simple, alternate leaves are 4 to 12 inches long. Flowers are greenish-white to purplish and small; they appear on the drooping or erect stalk. The dark berries are purplish to black and are attached to the stalk by a short stem.

Distribution: Pokeweed grows in damp fields and woods from Maine to Minnesota, southern Ontario to southwest Quebec, south to the Gulf of Mexico, Florida, and Texas, and in Hawaii. It has been introduced in California where it is an occasional weed.

Toxic Part: The leaves and roots are poisonous. In some regions, the young sprouts and stems are boiled and eaten after discarding the cooking water; the cooked product may also be purchased commercially in cans. The mature berries are relatively nontoxic. Intoxications generally arise from eating uncooked leaves in salads or mistaking the roots for parsnips or horseradish.

Toxin: Phytolaccatoxin and related triterpenes.

Clinical Findings: After a delay of 2 to 3 hours, nausea, vomiting, abdominal cramping, and diarrhea occur. Effects may last for as long as 48 hours. An elevated white blood cell count may occur due to the presence in the plant of a mitogen and is generally of no consequence.

Management: Intravenous hydration, antiemetics, and electrolyte replacement may be necessary for patients with severe gastrointestinal effects, particularly in children. Consultation with a Poison Control Center should be considered. See "Poisoning by Plants with Gastrointestinal Toxins," p. 28.

References

Hamilton RJ, Shih RD, Hoffman RS. Mobitz type I heart block after pokeweed ingestion. Vet Hum Toxicol 1995;1:66–67.

Roberge R, Brader E, Martin ML, et al. The root of evil-pokeweed intoxication. Ann Emerg Med 1986;15:470–473.

Pieris species
Family: Ericaceae
Pieris floribunda (Pursh) Benth. & Hook. f.
Pieris japonica (Thunb.) D. Don

Common Names: Fetter-bush, **Lily-of-the-Valley Bush**, Japanese Pieris

Description: *Pieris* are evergreen shrubs or small trees. The leaves are simple, usually alternate, and leathery. The flowers are usually white and grow in urn-shaped clusters.

Distribution: *Pieris floribunda* is a bush that grows from Virginia to Georgia. *Pieris japonica*, a shrub or small tree from Japan, is widely cultivated in temperate climates, particularly on the Pacific coast. Numerous horticultural varieties exist.

Pieris floribunda, close-up of flowers (above)

Pieris japonica (below)

Toxic Part: The leaves and nectar from flowers (in honey).

Toxin: Grayanotoxins (andromedotoxins), sodium channel activators.

Clinical Findings: Symptoms are predominantly neurological and cardiac. There is

Pieris japonica, flowering shrub

transient burning in the mouth after ingestion, followed after several hours by increased salivation, vomiting, diarrhea, and a tingling sensation in the skin. The patient may complain of headache, muscular weakness, and dimness of vision. Bradycardia and other cardiac dysrhythmias can be associated with severe blood pressure abnormalities. Coma may develop, and convulsions may be a terminal event.

Management: Fluid replacement should be instituted with respiratory support if indicated. Heart rhythm and blood pressure should be monitored and treated with appropriate medications and supportive care. Recovery is generally complete within 24 hours. Consultation with a Poison Control Center should be strongly considered. See "Poisoning by Plants with Sodium Channel Activators," p. 32.

References

Ergun K, Tufekcioglu O, Aras D, Korkmaz S, Pehlivan S. A rare cause of atrioventricular block: Mad Honey intoxication. Int J Cardiol 2005;99:347–348.

Puschner B, Holstege DM, Lamberski N. Grayanotoxin poisoning in three goats. J Am Vet Med Assoc 2001;218:573–575.

Tschekunow H, Klug S, Marcus S. Symptomatic poisoning from ingestion of *Pieris japonica*. A case report. Vet Hum Toxicol 1989;31:360.

Yavuz H, Ozel A, Akkus I, Erkul I. Honey poisoning in Turkey. Lancet 1991;337:789–790.

Podophyllum peltatum L.
Family: Berberidaceae

Common Names: American Mandrake, Behen, Devil's Apple, Hog Apple, Indian Apple, **May Apple**, Raccoon Berry, Umbrella Leaf, Wild Jalap, Wild Lemon

Description: This herb grows to 1.5 feet and has two large (up to 1 foot across) umbrella-like leaves, each with five to nine lobes. Between the leaf stems emerges a single, white, nodding flower with a 2-inch diameter. The yellowish-green fruit is about the size and shape of an egg. Sterile (nonflowering) plants have a single leaf.

Distribution: May apples grow in moist woodlands from Quebec to Florida, west to southern Ontario, Minnesota, and Texas.

Toxic Part: The whole plant, except the fruit (which causes, at most, only slight catharsis), is poisonous.

Podophyllum peltatum (above)

Podophyllum peltatum, close-up of flower (left)

Toxin: Podophyllotoxin, and alpha- and beta-peltatin, similar to colchicine, a cytotoxic alkaloid capable of inhibiting microtubule formation. Available for clinical use as podophyllin resin.

Clinical Findings: Ingestion of podophyllin resin may cause initial oropharyngeal pain followed, in several hours, by intense gastrointestinal effects. Abdominal pain and severe, profuse, persistent diarrhea may develop, with the latter causing extensive fluid depletion and electrolyte abnormalities. Podophyllin may subsequently produce peripheral neuropathy, bone marrow suppression, and cardiovascular collapse. Topical application may lead to skin necrosis or systemic poisoning.

Management: The intoxication has a prolonged course due to the slow absorption of the toxin. Aggressive symptomatic and supportive care is critical, with prolonged observation of symptomatic patients. Consultation with a Poison Control Center should be strongly considered. See "Poisoning by Plants with Mitotic Inhibitors," p. 29.

References

Cassidy DE, Drewry J, Fanning JP. *Podophyllum* toxicity: A report of a fatal case and a review of the literature. J Toxicol Clin Toxicol 1982;19:35–44.

Damayanthi Y, Lown JW. Podophyllotoxins: current status and recent developments. Curr Med Chem 1998;5:205–252.

Frasca T, Brett AS, Yoo SD. Mandrake toxicity. A case of mistaken identity. Arch Intern Med 1997;157:2007–2009.

Givens ML, Rivera W. Genital burns from home use of podophyllin. Burns 2005;31:394–395.

Schwartz J, Norton SA. Useful plants of dermatology. VI. The mayapple (*Podophyllum*). J Am Acad Dermatol 2002;47:774–775.

Stoehr GP, Peterson AL, Taylor NJ. Systemic complications of local podophyllin therapy, Ann Intern Med 1978;89:362–363.

Poncirus trifoliata (L.) Raf.
Family: Rutaceae

Common Names: Bitter Orange, Hardy Orange, **Mock Orange**, Trifoliate Orange

Description: This small deciduous tree has green stems with stiff thorns that may be 2.5 inches long. The white flowers are 2 inches across and form at the base of the thorns before leafing. Leaves are shiny, thick, and leathery with three leaflets. The fruit resembles a small orange and has a pronounced aroma. It is filled with an acrid pulp.

Distribution: *Poncirus* is used in the southern United States and along both coasts as a hedge plant. It also is cultivated extensively to obtain rootstock for citrus grafts.

Toxic Part: The fruit is toxic.

Toxin: Acridone alkaloid, a gastrointestinal toxin.

Clinical Findings: There are no adequately documented human poisonings, and clinical descriptions are based on the nature of the toxin. Serious intoxications are unlikely because of the bitter taste of the fruit; nausea, vomiting, abdominal cramping, and diarrhea may occur.

Poncirus trifoliata, close-up of flowers (right)

Poncirus trifoliata, tree showing fruit (below)

Management: Intravenous hydration, antiemetics, and electrolyte replacement may be necessary for patients with severe gastrointestinal symptoms, particularly in children. Consultation with a Poison Control Center should be considered. See "Poisoning by Plants with Gastrointestinal Toxins," p. 28.

Reference

Lee HT, Seo EK, Chung SJ, Shim CK. Prokinetic activity of an aqueous extract from dried immature fruit of Poncirus trifoliata (L.) Raf. J Ethnopharmacol 2005;102:131–136.

Prunus species
Family: Rosaceae

Common Names: Apricot, Cherry, Choke Cherry, Peach, Plum, Sloe

Description: *Prunus* species are trees and shrubs, usually with deciduous leaves that are alternate and mostly serrate. The flowers are white or pink with five sepals and five petals. The fruit is a drupe with a fleshy outer layer over a stone or pit.

Prunus pendula var. adscendens, flowering tree (above)

Distribution: *Prunus* is widely cultivated in the north temperate zone, and there are a number of native species. A large number of fruit varieties are available commercially.

Prunus pendula var. adscendens, flowers (below)

Toxic Part: The kernel in the pit is poisonous. Most fatal intoxications involve ingestion of apricot pits or products derived from them.

Toxin: Amygdalin, a cyanogenic glycoside.

Prunus spp., assorted fruits (above)

Prunus laurocerasus, branch with flowers (right)

Prunus serotina, branch with flowers

Clinical Findings: Because the cyanogenic glycosides must be hydrolyzed in the gastrointestinal tract before cyanide ion is released, several hours may elapse before poisoning occurs. Abdominal pain, vomiting, lethargy, and sweating typically occur first. Cyanosis does not occur. In severe poisonings, coma develops and may be accompanied by convulsions and cardiovascular collapse.

Management: Symptomatic and supportive care should be given. Antidotal therapy is available. Consultation with a Poison Control Center is strongly suggested. See "Poisoning by Plants with Cyanogenic Compounds," p. 27.

References

Pentore R, Venneri A, Nichelli P. Accidental choke-cherry poisoning: early symptoms and neurological sequelae of an unusual case of cyanide intoxication. Ital J Neurol Sci 1996; 17:233–235.

Sayre JW, Kaymakcalan S. Hazards to health: Cyanide poisoning from apricot seeds among children in central Turkey. N Engl J Med 1964;270:1113–1115.

Pteridium aquilinum (L.) Kuhn
Family: Dennstaedtiaceae

Common Names: Bracken Fern, Brake, Hog Pasture Break, Pasture Break

Description: This is an aggressively spreading, very common fern that covers large areas. The fronds are 2 to 4 feet tall, triangular, divided into three main parts, with coarse pinnae that are wooly on their underside. The fronds are borne on erect stalks.

Distribution: Bracken Fern is found in many areas of the United States, particularly on barren or abandoned land, and in tropical and temperate areas of the world. It can be found in cultivation as well.

Toxic Part: Entire plant. "Fiddleheads," the tender unfolding leaves of this species, are prepared like asparagus.

Toxin: Perhaps ptaquiloside, a sesquiterpene.

Clinical Findings: Acute ingestion of bracken fern by cattle may cause a potentially fatal syndrome characterized by fever, lethargy, and bleeding. Human toxicity has not been reported. However, ingestion of large amounts would likely

Pteridium aquilinum, "fiddleheads" (unfolding leaves) (left)

Pteridium aquilinum (below)

produce toxicity. Animal studies suggest that chronic ingestion of Bracken Fern can lead to intestinal and bladder tumors. No report of human tumor development is noted; however, they are thought to be related to the high incidence of stomach cancer in Japan, where they are commonly eaten.

Management: Symptomatic and supportive care. Consultation with a Poison Control Center should be considered.

References

Alonso-Amelot ME, Avendano M. Human carcinogenesis and bracken fern: A review of the evidence. Curr Med Chem 2002;9:675–686.

Marliere CA, Wathern P, Castro MC, O'Connor, Galvao MA. Bracken fern (*Pteridium aquilinum*) ingestion and oesophageal and stomach cancer. IARC Sci Publ 2002;156: 379–380.

Ranunculus species
Family: Ranunculaceae
Ranunculus acris L.
Ranunculus bulbosus L.
Ranunculus sceleratus L.

Ranunculus acris, flowers

Common Names: Bassinet, Blister Flower, Blister Wort, Bouton d'Or, Butter-Cress, **Buttercup**, Butter Daisy, Butter Flower, Crain, Crowfoot, Devil's Claws, Figwort, Goldballs, Goldweed, Horse Gold, Hunger Weed, Lesser Celandine, Pilewort, Ram's Claws, Renoncule, St. Anthony's Turnip, Sitfast, Spearwort, Starve-Acre, Water Crowfoot, Yellow Gowan

Description: These annual or perennial herbs have large basal leaves and yellow, white, or, rarely, red flowers. They are a few inches to under 3 feet in height.

Distribution: Buttercups grow mainly in wet, swampy

areas throughout the United States, including Hawaii, Alaska, and Canada.

Toxic Part: The sap is toxic. Poisoning has occurred in children who ate the bulbous roots of *Ranunculus bulbosus*.

Toxin: Protoanemonin, an irritant.

Clinical Findings: There are no adequately documented human poisonings, and clinical descriptions are derived primarily from animal reports. Intense pain and inflammation of the mouth with blistering, ulceration, and profuse salivation can occur. Bloody emesis and diarrhea develop in association with severe abdominal cramps. Central nervous system involvement is manifested by dizziness, syncope, and seizures.

Ranunculus sceleratus

Management: Most exposures result in minimal or no toxicity. Intravenous hydration, antiemetics, and electrolyte replacement may be necessary for patients with severe gastrointestinal effects, particularly in children. Consultation with a Poison Control Center should be considered. See "Poisoning by Plants with Gastrointestinal Toxins," p. 28. If seizures occur, rapidly acting anticonvulsants, such as intravenous diazepam, should be utilized along with other supportive measures. Consultation with a Poison Control Center should be strongly considered. See "Poisoning by Plants with Convulsive Poisons (Seizures)," p. 25.

Reference

Turner NJ. Counter-irritant and other medicinal uses of plants in Ranunculaceae by native peoples in British Columbia and neighbouring areas. J Ethnopharmacol 1984;11:181–201.

Rhamnus species
Family: Rhamnaceae
Rhamnus californica Eschsch.
Rhamnus cathartica L.
Rhamnus frangula L.
Rhamnus purshiana DC.

Common Names: Alder Buckthorn, Arrow Wood, Bearberry, Berry Alder, Black Dogwood, **Buckthorn**, Cáscara, Cáscara Sagrada, Hart's Horn, May Thorn, Nerprun, Persian Berry, Purging Buckthorn, Rhine Berry
Rhamnus californica: **Coffeeberry**

Rhamnus cathartica, fruiting branch (above)

Rhamnus cathartica, fruiting branch (below)

Description: Buckthorns are shrubs or small trees, 6 to 12 feet tall. The flowers are small and greenish or greenish-white. The fruits are "drupaceous," that is, they contain two or more hard "stones" or "pits," each enclosing a single seed. *Rhamnus californica* is an evergreen with finely toothed leaves. The flowers form in a flat-topped cluster. The fruit is red at first, turning to black when mature. *Rhamnus cathartica* has spine-tipped branchlets; opposite, toothed leaves; scaly buds; and greenish-white flowers that grow in clusters. The mature fruit is black and contains an even number of stones, usually four. *Rhamnus frangula* does not have spines and, unlike the preceding two species, its leaves have smooth margins (i.e., not toothed) with occasional glands. The buds of *R. frangula* are hairy. The flowers grow in a flat cluster. The fruit is red when young, black when mature, and contains two to three stones. *Rhamnus purshiana* is larger than others discusssed here (grows to 20 feet). Leaves are finely toothed; flowers grow in flat-topped clusters; mature fruits are black.

Distribution: *Rhamnus cathartica* and *R. frangula* have been introduced from Europe and are naturalized in the northeastern United States and Canada. *Rhamnus californica* grows in California. Related species are found throughout the north temperate zone.

Toxic Part: Fruit and bark are poisonous. However, the fruits of *Rhamnus cathartica* have long been used as a laxative, hence its species name. The bark of *Rhamnus purshiana*, a plant of the western United States, also produces a very strong laxative, known as cascara sagrada.

Toxin: Hydroxymethylanthraquinone, a gastrointestinal irritant.

Clinical Findings: Nausea, vomiting, abdominal cramping, and diarrhea may occur.

Management: Intravenous hydration, antiemetics, and electrolyte replacement may be necessary for patients with severe gastrointestinal effects, particularly in children. Consultation with a Poison Control Center should be considered. See "Poisoning by Plants with Gastrointestinal Toxins," p. 28.

References

Giavina-Bianchi PF Jr, Castro FF, Machado ML, Duarte AJ. Occupational respiratory allergic disease induced by *Passiflora alata* and *Rhamnus purshiana*. Ann Allergy Asthma Immunol 1997;79:449–454.

Lichtensteiger CA, Johnston NA, Beasley VR. *Rhamnus cathartica* (buckthorn) hepatocellular toxicity in mice. Toxicol Pathol 1997;25:449–452.

van Gorkom BA, de Vries EG, Karrenbeld A, Kleibeuker JH. Review article: Anthranoid laxatives and their potential carcinogenic effects. Aliment Pharmacol Ther 1999;13: 443–452.

Rheum × cultorum Thorsrud & Reisaeter
(=***R. rhabarbarum* L.)** (Note: this plant is an ancient hybrid probably involving *Rheum rhaponticum* L.)
Family: Polygonaceae

Common Names: Pie Plant, **Rhubarb**, Rhubarbe, Wine Plant

Description: This perennial has leaf stalks that grow several feet long and become red at maturity. The large ovate leaves are wrinkled with wavy margins, and the plant produces white flowers in a tall terminal cluster.

Rheum × cultorum

Distribution: Rhubarb is widely cultivated for its edible leaf stalk.

Toxic Part: Raw leaf blades are toxic if eaten in large quantity.

Toxin: Anthraquinone glycosides and soluble oxalates.

Clinical Findings: Toxicity often results when novices mistakenly prepare the leaf blades instead of the edible leaf stalks. Effects are caused by the cathartic effect of anthraquinone glycosides. Nausea, vomiting, abdominal cramping, and diarrhea may occur. Ingestion of a sufficient amount of the leaf can lead to systemic hypocalcemia, caused by absorption of soluble oxalates with subsequent precipitation of calcium oxalate crystals in body fluids. In the urine, this effect may produce renal damage.

Management: Intravenous hydration, antiemetics, and electrolyte replacement (particularly calcium and potassium) may be necessary. Close monitoring of urine output and renal function is important to detect renal injury. Consultation with a Poison Control Center should be considered. See "Poisoning by Plants with Gastrointestinal Toxins," p. 28.

References

Jacobziner H, Raybin HW. Rhubarb poisoning. NY State J Med 1962;62:1676–1678.

Sanz P, Reig R. Clinical and pathological findings in fatal plant oxalosis. A review. Am J Forensic Med Pathol 1992;13:342–345.

Rhododendron species
Family: Ericaceae

Common Names: Azalea, **Rhododendron**, Rhodora, Rosa Laurel, Rosebay

Description: There are about 800 species of *Rhododendron*, divided into eight subgenera. The genus includes the former genus *Azalea*. Rhododendrons are evergreen, semievergreen, or deciduous shrubs with simple, alternate leaves. The flowers are of various colors and bell shaped or funnel-form.

Distribution: Rhododendrons are cultivated in Canada and most of the United States with the exception of the north central states and subtropical Florida. A number of native species exist in the same geographic range.

Toxic Part: The leaves are toxic, as is honey made from flower nectar.

Toxin: Grayanotoxins (andromedotoxins), sodium channel activators.

Clinical Findings: Symptoms are predominantly neurological and cardiac. There is transient burning in the mouth after ingestion, followed after several hours by increased salivation, vomiting, diarrhea, and a tingling sensation in the skin. The patient may complain of headache, muscular weakness, and dimness of vision. Bradycardia and other cardiac dysrhythmias can be associated with severe blood pressure abnormalities. Coma may develop, and convulsions may be a terminal event.

Rhododendron cv. 'Yaku Princess,' branch with flowers (above)

Rhododendron cv. 'Rosebud,' flowers (below)

Rhododendron cv. 'PMJ,' branch with flowers

Management: Fluid replacement should be instituted with respiratory support if indicated. Heart rhythm and blood pressure should be monitored and treated with appropriate medications and supportive care. Recovery is generally complete within 24 hours. Consultation with a Poison Control Center should be strongly considered. See "Poisoning by Plants with Sodium Channel Activators," p. 32.

References

Ergun K, Tufekcioglu O, Aras D, Korkmaz S, Pehlivan S. A rare cause of atrioventricular block: Mad honey intoxication. Int J Cardiol 2005;99:347–348.

Puschner B, Holstege DM, Lamberski N. Grayanotoxin poisoning in three goats. J Am Vet Med Assoc 2001;218:573–575.

Sutlupinar N, Mat A, Satganoglu Y. Poisoning by toxic honey in Turkey. Arch Toxicol 1993:67;148–150.

von Malottki K, Wiechmann HW. Acute life-threatening bradycardia due to poisoning with Turkish wild honey. Dtsch Med Wochenschr 1996;121:936–938.

Yavuz H, Ozel A, Akkus I, Erkul I. Honey poisoning in Turkey. Lancet 1991;337: 789–790.

Rhodotypos scandens (Thunb.) Makino
(=*R. kerrioides* Siebold & Zucc.; *R. tetrapetala* (Sielbold) Makino)
Family: Rosaceae

Common Names: Jetbead, White Kerria

Description: This spreading, deciduous shrub grows 3 to 6 feet high. The leaves are opposite, toothed, and 2 inches long. The flowers are white and about 2

Rhodotypos scandens

Rhodotypos scandens, mature fruit

inches in diameter with four petals. The fruit is a shiny black drupe about 0.25 inch in diameter.

Distribution: Jetbead is hardy in the northern United States and is a popular ornamental.

Toxic Part: Fruits may be toxic.

Toxin: Not known (reputed to contain a cyanogenic glycoside).

Clinical Findings: In the only report of human poisoning, the patient exhibited severe hypoglycemia, ketosis, hyperthermia, and convulsions.

Management: The patient was treated with antipyretics and anticonvulsants. Recovery was complete by day 5. No specific antidote exists. If toxicity occurs, treatment is aimed at supportive measures. Consultation with a Poison Control Center should be strongly considered. See "Poisoning by Plants with Cyanogenic Compounds," p. 27.

Reference

Rascoff H, Wasser S. Poisoning in a child simulating diabetic coma. JAMA 1953:152:1134–1135.

Ricinus communis L.
Family: Euphorbiaceae

Common Names: African Coffee Tree, **Castor Bean**, Castor Oil Plant, Higuereta, Higuerilla, Koli, La'Au-'Aila, Man's Motherwort, Mexico Weed, Pa'Aila, Palma Christi, Ricin, Ricino, Steadfast, Wonder Tree

Description: This annual grows to 15 feet or higher in the tropics. The large, lobed leaves are up to 3 feet across. It is also grown as a summer ornamental in temperate areas where, depending on the cultivar, the leaves can be green to red-purple. Spiny fruits form in clusters along spikes. The fruits contain plump seeds resembling fat ticks in shape, usually mottled black or brown on white. The highly toxic seeds have a pleasant taste.

Distribution: The castor bean grows throughout the West Indies and is a naturalized weed in Florida, along the Gulf Coast in Texas, along the Atlantic Coast to New Jersey, in southern California, and in Hawaii. It is widely planted elsewhere for its foliage and is grown commercially in some of the Gulf Coast states and Guam.

Toxic Part: The toxin is contained within the hard, water-impermeable coat of the seeds. The toxin is not released unless the seed coats are broken (e.g., chewed) and the contents digested.

Ricinus communis

Ricinus communis

Ricinus communis, fruits

Ricinus communis, seeds

Toxin: Ricin.

Clinical Findings: Ingested seeds that remain intact as they pass through the gastrointestinal tract generally do not release toxin or cause toxicity. However, if the seeds are chewed, pulverized, or digested (i.e., if passage through the gastrointestinal tract is delayed), then the toxin is absorbed by intestinal cells causing mild to severe gastrointestinal toxicity. Effects depend upon the amount of toxin exposure and include nausea, vomiting, abdominal cramping, diarrhea, and dehydration. Variations in the severity of toxicity may be related to the degree to which the seeds are ground or chewed before ingestion. Parenteral administration (such as by injection or inhalation), or perhaps a large ingestion, may produce life-threatening systemic findings, including multisystem organ failure, even with small exposures.

Management: Ingestion of intact seeds does not cause toxicity in the majority of cases and requires no therapy. Cases associated with gastrointestinal effects need to be assessed for signs of dehydration and electrolyte abnormalities. Activated charcoal should be administered. Intravenous hydration, antiemetics, and electrolyte replacement may be necessary in severe cases, particularly in children. Consultation with a Poison Control Center should be strongly considered. See "Poisoning by Plants with Toxalbumins," p. 33.

References

Aplin PJ, Eliseo T. Ingestion of castor oil plant seeds. Med J Aust 1997;167:260–261.

Audi J, Belson M, Patel M, Schier J, Osterlob J. Ricin poisoning: A comprehensive review. JAMA 2005;294:2342–2351.

Bradberry SM, Dickers KJ, Rice P, Griffiths GD, Vale JA. Ricin poisoning. Toxicol Rev 2003;22:65–70.

Doan LG. Ricin: mechanism of toxicity, clinical manifestations, and vaccine development. A review. J Toxicol Clin Toxicol 2004;42:201–208.

Hartley MR, Lord JM. Cytotoxic ribosome-inactivating lectins from plants. Biochim Biophys Acta 2004;1701:1–14.

Metz G, Bocher D, Metz J. IgE-mediated allergy to castor bean dust in a landscape gardener. Contact Dermatitis 2001;44:367.

Olsnes S, Kozlov JV. Ricin. Toxicon 2001;39:1723–1728.

Rivina humilis L.
Family: Phytolaccaceae

Common Names: Baby Pepper, Bloodberry, Caimonicillo, Carmín, Cat's Blood, Coral Berry, Coralitos, Pigeon Berry, **Rouge Plant**

Description: This small shrub grows to 4 feet. The leaves are ovate and about 4 inches long. The greenish to pinkish flowers grow in hanging sprays up to 8 inches in length. The fruit is a shiny bright crimson or orange berry.

Rivina humilis, mature fruits on stems

Distribution: Rouge plant grows in Hawaii, southern United States from New Mexico to Florida, and the West Indies. It is a popular houseplant elsewhere.

Toxic Part: The leaves and roots are poisonous. The absence of reports of poisoning from the attractive berries suggests that they do not contain clinically significant concentrations of toxin.

Toxin: An unknown toxin similar to that in *Phytolacca americana.*

Clinical Findings: Nausea, vomiting, abdominal cramping, and diarrhea may occur.

Management: Intravenous hydration, antiemetics, and electrolyte replacement may be necessary for patients with severe gastrointestinal effects, particularly in children. Consultation with a Poison Control Center should be considered. See "Poisoning by Plants with Gastrointestinal Toxins," p. 28.

Reference

Smith GW, Constable PD. Suspected pokeweed toxicity in a boer goat. Vet Hum Toxicol 2002;44:351–353.

Robinia pseudoacacia L.
Family: Leguminosae (Fabaceae)

Common Names: Bastard Acacia, Black Acacia, **Black Locust**, False Acacia, Green Locust, Pea Flower Locust, Post Locust, Silver Chain, Treesail, White Honey Flower, White Locust, Yellow Locust

Description: The locust tree grows to 80 feet. The alternate compound leaves have 7 to 19 elliptical leaflets, 1 inch in length. There is a pair of woody spines on the branch where the leaf stem emerges. The white, fragrant flowers are borne in drooping clusters (racemes). The fruit is a flat, reddish-brown pod, about 4 inches long, which persists over the winter.

Distribution: The locust is native to the Appalachian Mountains and the Ozarks. It is widely planted in the north temperate zones, particularly in the east, and in Ontario east to Nova Scotia and southern British Columbia.

Toxic Part: The bark, seeds, and leaves are toxic.

Toxin: Robin, a plant lectin (toxalbumin) related to ricin.

Clinical Findings: Poisoning from this tree is potentially serious, but exceedingly uncommon. Effects depend upon the amount of toxin exposure and include nausea, vomiting, abdominal cramping, diarrhea, and dehydration. Variations in the severity of toxicity may be related to the degree to which the seeds are ground or chewed before ingestion. Parenteral administration (such as by injection or inhalation), or perhaps a large ingestion, may produce life-threatening systemic findings, including multisystem organ failure, even with small exposures.

Robinia pseudoacacia, flowering tree (above)

Robinia pseudoacacia, close-up of flowers (below)

Management: Cases associated with gastrointestinal effects need to be assessed for signs of dehydration and electrolyte abnormalities. Activated charcoal should be administered. Intravenous hydration, antiemetics, and electrolyte replacement may be necessary in severe cases, particularly those involving children. Consultation with a Poison Control Center should be strongly considered. See "Poisoning by Plants with Toxalbumins," p. 33.

References

Artero Sivera A, Arnedo Pena A, Pastor Cubo A. Clinico-epidemiologic study of accidental poisoning with *Robinia pseudoacacia* L. in school children. An Esp Pediatr 1989;30: 191–194.

Costa Bou X, Soler i Ros JM, Seculi Palacios JL. Poisoning by *Robinia pseudoacacia*. An Esp Pediatr 1990;32:68–69.

Hui A, Marraffa JM, Stork CM. A rare ingestion of the Black Locust tree. J Toxicol Clin Toxicol 2004;42:93–95.

Mejia MJ, Morales MM, Llopis A, Martinez I. School children poisoned by ornamental trees. Aten Primaria 1991;8:88, 90–91.

Sambucus species
Family: Caprifoliaceae
Sambucus nigra L.
Sambucus nigra L. ssp. *caerulea* (Raf.) R. Bolli (=*S. caerulea* Raf.)
Sambucus nigra L. ssp. *canadensis* (L.) R. Bolli (=*S. canadensis* L.; *S. simpsonii* Rehder; *S. mexicana* C. Presl ex DC.)
Sambucus melanocarpa A. Gray
Sambucus racemosa L.

Common Names: American Elder, Dauco, **Elderberry**, Fleur Sureau, Saúco Blanco, Surreau

Description: The common elderberry (*Sambucus nigra* ssp. *canadensis*) in the eastern half of the United States and Canada is a 6- to 12-foot shrub with much branched stems. The compound leaves usually have seven serrated oval leaflets. The small, white flowers are borne in flat-topped, terminal clusters. The purplish-black fruits are about 0.35 inch in diameter. Flowers and fruit may be borne simultaneously.

Sambucus nigra, close-up of fruit

Distribution: Species of elderberry may be found in all states including Hawaii and Alaska, across Canada, and in the Greater Antilles.

Toxic Part: The whole plant is poisonous, particularly the root. The ripe fruit is rendered harmless by cooking. *Sambucus nigra* is now used to make a popular herbal remedy, sold as a syrup or lozenge, that has been used to treat the influenza virus. The flowers are probably nontoxic.

Sambucus nigra ssp. *caerulea*, branch with fruit (above)

Sambucus nigra ssp. *canadensis*, close-up of flowers (left)

Toxin: Cyanogenic glycosides and/or toxalbumin.

Sambucus racemosa, branch with fruit

Clinical Findings: There are no adequately documented human cyanide poisonings from this plant genus, and clinical descriptions are based on the nature of the toxin. Because the cyanogenic glycosides must be hydrolyzed in the gastrointestinal tract before cyanide ion is released, several hours may elapse before poisoning occurs. Abdominal pain, vomiting, lethargy, and sweating typically occur first. Cyanosis does not occur. In severe poisonings, coma develops and may be accompanied by convulsions and cardiovascular collapse.

Toxalbumin ingestions typically cause mild to severe gastrointestinal toxicity, although parenteral administration (such as by injection or inhalation), or perhaps large ingestion, may produce life-threatening systemic findings, including multisystem organ failure, even with small exposures.

Management: Symptomatic and supportive care should be given. Antidotal therapy is available. Consultation with a Poison Control Center is strongly suggested. See "Poisoning by Plants with Cyanogenic Compounds," p. 27. and See "Poisoning by Plants with Toxalbumins," p. 33.

References

Battelli MG, Citores L, Buonamici L, et al. Toxicity and cytotoxicity of nigrin b, a two-chain ribosome-inactivating protein from *Sambucus nigra*: Comparison with ricin. Arch Toxicol 1997;71:360–364.

Buhrmester RA, Ebinger JE, Seigler DS. Sambunigrin and cyanogenic variability in populations of *Sambucus canadensis* L. (Caprifoliaceae). Biochem Syst Ecol 2000;28: 689–695.

Kunitz S, Melton RJ, Updyke T, Breedlove D, Werner SB. Poisoning from elderberry juice—California. MMWR (Morb Mortal Wkly Rep) 1984;33:173–174.

Sanguinaria canadensis L.
Family: Papaveraceae

Common Names: Bloodroot, Red Puccoon

Description: This perennial herb has a simple leaf that is palmately lobed, to about 6 inches across, and contains yellow sap in its stem. The flowers are white, sometimes with a tinge of pink, and on stems about 8 inches tall. The fruit is a capsule. The prominent rhizomes contain red sap and are dried for medicinal use.

Distribution: Found in the deep, rich woodlands of the eastern half of North America.

Sanguinaria canadensis, close-up of flower

Toxic Part: All parts of the plant contain alkaloids that are similar to ones derived from the related opium poppy. However, these alkaloids are most concentrated in the root.

Toxin: Sanguinarine, chelerythrine, protopine, and homochelidonine, all isoquinoline alkaloids. It is not clear which of these is the toxic component.

Clinical Findings: Bloodroot is a traditional herbal

remedy utilized by Native Americans of North America. It continues to be used as a medicinal agent and an oral dentrifice. Used this way, there are few reports of toxicity. However, because of its similarity to the opium poppy, patients with substantial ingestion should be observed for altered mental status and respiratory depression.

Management: Patients that manifest mental status alteration may theoretically benefit from the use of naloxone. However, no case reports document the efficacy of this agent. In addition, symptomatic and supportive care should be instituted. Consultation with a Poison Control Center should be considered.

References

Becci PJ, Schwartz H, Barnes HH, Southard GL. Short-term toxicity studies of sanguinarine and of two alkaloid extracts of *Sanguinaria canadensis* L. J Toxicol Environ Health 1987; 20:199–208.

Kosina P, Walterova D, Ulrichova J, et al. Sanguinarine and chelerythrine: Assessment of safety on pigs in ninety days feeding experiment. Food Chem Toxicol 2004;42:85–91.

Sapindus species
Family: Sapindaceae
Sapindus saponaria L. (=*S. marginatus* Willd.)
Sapindus drummondii Hook. & Arn.

Common Names:
Sapindus drummondii: **Western Soapberry**
Sapindus saponaria: A'e, Bois Savonnette, False Dogwood, Indian Soap Plant, Jaboncillo, Manele, Savonnier, **Soapberry**, Wild China Tree, Wingleaf Soapberry

Description: *Sapindus drumondii*: A deciduous tree growing to 50 feet tall with pinnate leaves containing 8 to 10 leaflets, each about 3 inches long. Flowers are small, yellowish-white, in panicles. Fruits are yellow, turning black, to 0.5 inch long. *Sapindus saponaria*: A tropical evergreen tree growing to 30 feet. Leaves with 7 to 9 leaflets, each to 4 inches long. Flowers are white, and fruits are shiny orange-brown, about 0.75 inch in diameter. The fruit of these species has been employed as soap.

Distribution: *Sapindus drummondii* grows from Arizona to Louisiana north to Kansas. *Sapindus saponaria* grows in Florida, the West Indies, Mexico, and Hawaii.

Toxic Part: The fruit is poisonous.

Sapindus saponaria, fruiting branch

Toxin: Saponin, a gastrointestinal irritant, and a dermal irritant/sensitizer.

Clinical Findings: Most ingestions result in little or no toxicity. The saponins are poorly absorbed, but with large exposures gastrointestinal effects of nausea, vomiting, abdominal cramping, and diarrhea may occur. Allergic sensitization to this plant is common and can cause severe dermatitis.

Management: If severe gastrointestinal effects occur, intravenous hydration, antiemetics, and electrolyte replacement may be necessary, particularly in children. Consultation with a Poison Control Center should be considered. See "Poisoning by Plants with Gastrointestinal Toxins," p. 28.

Reference

Morton JF. Ornamental plants with poisonous properties. II. Proc Fla Hortic Soc 1962; 75:484.

Schinus species
Family: Anacardiaceae
Schinus molle L.
Schinus terebinthifolius Raddi

Common Names:
Schinus molle: Árbol del Perú, California Pepper Tree, **Pepper Tree**, Peruvian Mastic Tree, Pimiento de América
Schinus terebinthifolius: Brazilian Pepper Tree, **Christmas Berry Tree**, Copal, Florida Holly, Nani-O-Hilo, Pimienta del Brazil, Pink Pepper, Wilelaiki

Schinus molle (above)

Schinus molle, fruiting branch (left)

Description:

Schinus molle: This tree grows to a height of about 35 feet and often has a twisted trunk. When viewed from a distance, the compound leaves and hanging branches are markedly similar to the willow. *Schinus* bears small red fruits. The crushed leaves emit a distinct odor of black pepper.

Schinus terebinthifolius: This fast-growing tree or

Schinus terebinthifolius, fruiting branch

shrub usually attains a height of about 20 feet. The leaves are compound; individual leaflets are about 2.5 inches long. The leaves emit a pepper-like odor when crushed, but it is not as pronounced as with *S. molle*. The fruit forms in persistent clusters.

Distribution: *Schinus molle* is cultivated in California, Hawaii, and the West Indies. *Schinus terebinthifolius*, an introduced weed tree, is becoming common in south Florida, Hawaii, Guam, and the West Indies.

Toxic Part: Fruits are poisonous. Sometimes they are eaten when mixed with white or black peppercorns (both from *Piper nigrum*) to which they have been added for color.

Toxin: Triterpenes are gastrointestinal irritants.

Clinical Findings: Skin contact can cause contact dermatitis. Nausea, vomiting, abdominal cramping, and diarrhea may occur.

Management: Intravenous hydration, antiemetics, and electrolyte replacement may be necessary for patients with severe gastrointestinal effects, particularly in children. Consultation with a Poison Control Center should be considered. See "Poisoning by Plants with Gastrointestinal Toxins," p. 28.

References

Campello J de P, Marsaioli AJ. Triterpenes of *Schinus terebenthefolius*. Phytochemistry 1974;13:659–660.

Stahl E, Keller K, Blinn C. Cardanol, a skin irritant in pink pepper. Planta Med 1983;48: P5–P9.

Schoenocaulon species
Family: Liliaceae
Schoenocaulon drummondii A. Gray
Schoenocaulon texanum Scheele

Common Names:
Schoenocaulon drummondii: **Green Lily**
Schoenocaulon texanum: **Texas Green Lily**

Description: These two species are similar, and their botanical names once were considered to be synonyms. They are perennial herbs growing from bulbs. The leaves are grasslike, emerging only at the base. The flowering stalk is leafless. Flowers grow in a dense raceme and are pale green to yellowish-white. *Schoenocaulon drummondii* usually flowers in the fall, *S. texanum* during spring and summer. The fruit is a persistent, three-lobed capsule; each capsule contains four or more seeds.

Distribution: *Schoenocaulon* is a New World genus of about 10 species distributed mostly in Mexico south to Peru. *Schoenocaulon drummondii* is found in the central coastal prarie of Texas to the central Rio Grande Plains, and northeastern Mexico; *S. texanum* is found from the northeastern Rio Grande Plains westward through the Trans-Pecos and southeastern part of New Mexico and northern Mexico.

Toxic Part: The whole plant, particularly the seeds, is toxic.

Toxin: Veratrum alkaloids (steroidal alkaloids), sodium channel activators; and unidentified gastrointestinal toxins.

Clinical Findings: Reports of poisoning are infrequent. Symptoms are predominantly neurological and cardiac. There is transient burning in the mouth after ingestion, followed after several hours by increased salivation, vomiting, diarrhea, and a tingling sensation in the skin. The patient may complain of headache, muscular weakness, and dimness of vision. Bradycardia and other cardiac dysrhythmias can be associated with severe blood pressure abnormalities. Coma may develop, and convulsions may be a terminal event.

Management: Fluid replacement should be instituted with respiratory support if indicated. Heart rhythm and

Schoenocaulon drummondii, close-up of inflorescence

blood pressure should be monitored and treated with appropriate medications and supportive care. Recovery is generally complete within 24 hours. Consultation with a Poison Control Center should be strongly considered. See "Poisoning by Plants with Sodium Channel Activators," p. 32.

References

Grancai D, Grancaiova Z. Veratrum alkaloids I. Ceska Slov Farm 1994;43:147–154.

Gaillard Y, Pepin G. LC-EI-MS determination of Veratridine and Cevadine in two fatal cases of veratrum album poisoning. J Anal Toxicol 2001;25:481–485.

Scilla species
Family: Liliaceae

Common Names: Cuban Lily, Hyacinth-of-Peru, Jacinto de Perú, Peruvian Jacinth, Sea Onion, **Squill**, Star Hyacinth

Description: This plant arises from a bulb, has straplike leaves, and grows 6 to 12 inches tall. The flowers of this hyacinth-like plant are usually blue, purple, or white.

Distribution: These European or Asian plants are hardy perennials in the north temperate zones to British Columbia and Quebec to Newfoundland. They are often cultivated.

Toxic Part: The whole plant is poisonous.

Toxin: Cardioactive steroids resembling digitalis.

Clinical Findings: There are no adequately documented human poisonings, and clinical descriptions are derived primarily from animal reports. Substantial ingestion may lead to toxicity. Poisoning would be expected to produce Clinical Findings typical of cardioactive steroid poisoning. Toxicity has a variable latent period that depends on the quantity ingested. Dysrhythmias are usually expressed as sinus bradycardia, premature ventricular contractions, atrioventricular conduction defects, or ventricular tachydysrhythmias. Hyperkalemia, if present, may be an indicator of toxicity.

Management: Gastrointestinal decontamination as appropriate, serial electrocardiograms, and serum potassium determinations should be performed. If serious cardioactive steroid toxicity is considered, digoxin-specific Fab should

Scilla peruviana (right)

Scilla hispanica (below)

Scilla siberica (above)

Scilla sinensis (left)

be administered. Consultation with a Poison Control Center should be considered. See "Poisoning by Plants with Cardioactive Steroids/Cardiac Glycosides," p. 24.

References

Aliotta G, De Santo NG, Pollio A, Sepe J, Touwaide A. The diuretic use of *Scilla* from Dioscorides to the end of the 18th century. J Nephrol 2004;17:342–347.

Verbiscar AJ, Patel J, Banigan TF, Schatz RA. Scilliroside and other *Scilla* compounds in red squill. J Agric Food Chem 1986;34:973–979.

Von Wartburg A, Kuhn M, Huber K. Cardiac glycosides from the white sea onion or squill. The constitution of the scilliphaeosides and glucoscilliphaeosides. Helv Chim Acta 1968; 51:1317–1328.

Senecio species
Family: Compositae (Asteraceae)
Senecio douglasii DC. var. *longilobus* (Benth.) L. Benson
Senecio jacobaea L.
Senecio vulgaris L.

Common Names:
Senecio jacobaea: Hierba de Santiago, **Ragwort**, Stinking Willie, Tansy Ragwort
Senecio douglasii var. *longilobus*: **Threadleaf Groundsel**
Senecio vulgaris: Common Groundsel, **Groundsel**

Description: There are 2,000 to 3,000 species of *Senecio*, many of which contain a toxic concentration of pyrrolizidine alkaloids.

Senecio douglasii

Senecio jacobaea: This biennial or perennial herb grows to 4 feet. It has yellow flowers and a shrubby form with many branching stems arising from the base. The new stems and leaves have a cottony covering.

Senecio douglasii var. *longilobus*: This very showy perennial has yellow flowers. Similar to *S. jacobaea*, its growth form is shrubby, with many branching stems arising from the base, and its new stems and leaves have a cottony covering.

Senecio vulgaris: This annual grows to about 1 foot high. Young stems are cottony but become smooth with aging. The leaves are soft, fleshy, and somewhat lobed. The flowers are golden-yellow.

Distribution: *Senecio jacobaea* is of Old World origin and has become naturalized as a weed in Newfoundland, Quebec, and Ontario south to Massachusetts; on the West Coast, it has become naturalized in British Columbia, Washington, and Oregon west of the Cascade Mountain range. *Senecio douglasii* var. *longilobus* grows in Colorado and Utah south to western Texas and northern Mexico. *Senecio vulgaris* is a European plant that has become naturalized as a weed in Alaska, all the Canadian provinces, and most of the continental United States.

Toxic Part: The whole plant is poisonous. Milk from animals who have grazed on these plants and honey made from nectar of *Senecio* contain the toxic alkaloids.

Toxin: Pyrrolizidine alkaloids.

Clinical Findings: Substantial short-term exposure may cause acute hepatitis, and chronic exposure to lower levels (including *Senecio* in herbal teas) may cause hepatic veno-occlusive disease (Budd–Chiari syndrome) and in some cases pulmonary hypertension.

Management: There is no known antidote. Supportive care is the mainstay of therapy. Consultation with a Poison Control Center should be considered. See "Poisoning by Plants with Pyrrolizidine Alkaloids," p. 31.

Senecio jacobaea

References

Fox DW, Hart MC, Bergeson PS, Jarrett PN, Stillman AE, Huxtable RJ. Pyrrolizidine (*Senecio*) intoxication mimicking Reye's syndrome. J Pediatr 1978;93:980–982.

Ortiz Cansado A, Crespo Valades E, Morales Blanco P, et al. Veno-occlusive liver disease due to intake of *Senecio vulgaris* tea. Gastroenterol Hepatol 1995;18:413–416.

Radal M, Bensaude RJ, Jonville-Bera AP, et al. Veno-occlusive disease following chronic ingestion of drugs containing *Senecio*. Therapie 1998;53:509–511.

Stegelmeier BL, Edgar JA, Colegate SM, et al. Pyrrolizidine alkaloid plants, metabolism and toxicity. J Nat Toxins 1999;8:95–116.

Stewart MJ, Steenkamp V. Pyrrolizidine poisoning: A neglected area in human toxicology. Ther Drug Monit 2001;23:698–708.

Sesbania species
Family: Leguminosae (Fabaceae)

Common Names: Báculo, Coffeeweed, Colorado River Hemp, Egyptian Rattlepod, Gallito, ʻOhai, ʻOhai-KeʻOkeʻO, ʻOhai-ʻUlaʻUla, Pois Valière, **Rattlebox**, Scarlet Wisteria Tree, Sesban, Vegetable Humming Bird

Sesbania grandiflora, leaves and flowers (above)

Sesbania grandiflora, branch with flowers and seed pods (right)

Description: These annuals have green stems 3 to 8 feet tall that become woody; the entire plant can be shrublike. The compound leaves have numerous linear leaflets. The small, sweetpea-shaped flowers are yellow dotted with purple. The fruits are curved seed pods.

Distribution: *Sesbania* grows in southern California, the south Atlantic and Gulf Coast states, Hawaii, Guam, and the West Indies.

Toxic Part: All parts of this plant are poisonous.

Toxin: Pyrrolizidine alkaloids.

Clinical Findings: There are no adequately documented human poisonings, and clinical descriptions are based on the nature of the toxin. Substantial short-term exposure may cause acute hepatitis, and chronic exposure to lower levels may cause hepatic veno-occlusive disease (Budd–Chiari syndrome) and in some cases pulmonary hypertension.

Management: There is no known antidote. Supportive care is the mainstay of therapy. Consultation with a Poison Control Center should be considered. See "Poisoning by Plants with Pyrrolizidine Alkaloids," p. 31.

Reference

Smith LW, Culvenor CCJ. Plant sources of hepatotoxic pyrrolizidine alkaloids. J Nat Prod 1981;44:129–152.

Solandra species
Family: Solanaceae

Common Names: Bejuco de Peo, **Chalice Vine**, Chamico Bejuco, Cup-of-Gold, Palo Guaco, Silver Cup, Trumpet Plant

Description: These climbing or erect tropical woody vines have large, showy, yellow or creamy-yellow trumpet-shaped flowers. The fruit is a fleshy, elongated berry.

Distribution: These plants are native to tropical America and Mexico. They are cultivated outdoors in Florida, the West Indies, and Hawaii.

Solandra grandiflora, flower

Solandra guttata, flowers (left)

Solandra guttata (below)

Toxic Part: All parts of this plant are toxic, including the flower nectar.

Toxin: Atropine, scopolamine, and other anticholinergic alkaloids.

Clinical Findings: Intoxication results in dry mouth with dysphagia and dysphonia, tachycardia, and urinary retention. Elevation of body temperature may be accompanied by flushed, dry skin. Mydriasis, blurred vision, excitement and delirium, headache, and confusion may be observed.

Management: Initially, symptomatic and supportive care should be given. If the severity of the intoxication warrants intervention (hyperthermia, delirium), an antidote, physostigmine, is available. Consultation with a Poison Control Center should be considered. See "Poisoning by Plants with Anticholinergic (Antimuscarinic) Poisons," p. 21.

Reference

Lozoya X, Meckes-Lozoya M, Chavez MA, Becerril G. Pharmacological and phytochemical study of 2 species of *Solandra* native to Mexico. Ars Pharm 1988;29:35–42.

Solanum species
Family: Solanaceae

Solanum americanum Mill., *S. nigrum* L., *S. ptychanthum* Dunal, and the *S. nigrum*-complex
Solanum capsicoides All. (=S. *aculeatissimum* sensu Britton & Millsp. non Jacq.; *S. ciliatum* Lam.)

Solanum americanum, flowers and young fruit (right)

Solanum americanum, close-up of fruit (below)

Solanum capsicoides, fruit (above)

Solanum capsicoides (left)

Solanum carolinense L.
Solanum dulcamara L.
Solanum linnaeanum
    Hepper & Jaeger
    (=*Solanum sodomeum*
    of authors not L.)
Solanum mammosum L.
Solanum pseudocapsicum L.
    (=*Solanum capsicastrum*
    Link ex Schauer)
Solanum seaforthianum
    Andrews
Solanum tuberosum L.

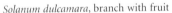

Solanum dulcamara, branch with fruit

Common Names:

Solanum americanum, S. nigrum, and the *S. nigrum* complex: Black Nightshade, Deadly Nightshade, Hierba Mora, Lanment, Mata Gallina, **Nightshade**, Poisonberry, Pop-Bush, Tue Chien, Yerba Mora

Solanum capsicoides: Berenjena de Jardín, Cockroach Berry, Kikinia-Lei, **Love Apple**, Pantomina, Soda-Apple Nightshade, Thorny Popolo

Solanum carolinense: Ball Nettle, Ball Nightshade, Bull Nettle, **Carolina Horse Nettle**, Sand Briar, Tread Softly

Solanum dulcamara, close-up of flowers (above)

Solanum mammosum, fruiting shrub (below)

Solanum mammosum, fruit cut in half

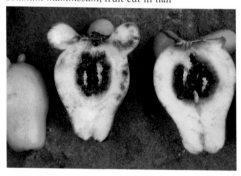

Solanum dulcamara: Agridulce, **Bittersweet**, Climbing Nightshade, Deadly Nightshade, Dog-Wood, Fellen, Felonwort, Morelle Douce-Amère, Poison Berry, Scarlet Berry, Snake Berry, Woody Nightshade

Solanum mammosum: Berenjena de Cucarachas, Berenjena de Marimbo, Berenjena de Gallina, Guirito, Love Apple, **Nipplefruit**, Pomme d'Amour, Tété Jeune Fille

Solanum pseudocapsicum: Christmas Orange, Coral, **Jerusalem Cherry**

Solanum seaforthianum: Brazilian Nightshade, Falsa Belladona, Jazmin de Italia, Lilas, **Star-Potato Vine**

Solanum sodomeum: **Apple of Sodom**, Dead Sea Apple, Popolo-Kikania, Thorny Popolo, Yellow Popolo

Solanum tuberosum: Papa, Patate, **Potato**, 'Uala-Kahiki

Description: *Solanum* is a very large genus with 1,700 species, most of which have not been evaluated toxicologically. These plants are mostly herbs (sometimes climbing) or shrubs. They are often spiny, hairy, or have stinging hairs. The flowers have five spreading petals, are often showy, and usually white or blue with five erect yellow stamens. The berries are black, orange, yellow, or red.

Solanum pseudocapsicum (above)

Solanum seaforthianum (left)

Distribution:

The *Solanum nigrum* complex includes several very similar and easily confused species, and the name *S. nigrum* is often applied to all of them. *Solanum ptychanthum* grows primarily in the eastern United States, Nova Scotia to Florida, west to North Dakota and Texas. *Solanum americanum* is found in tropical areas of the southern United States, the West Indies, Guam, and Hawaii. *Solanum nigrum* has been introduced from Europe and Asia to both coasts of the United States. *Solanum capsicoides* grows in Hawaii, on the coastal plain from Texas to North Carolina, and in the West Indies. *Solanum carolinense* grows from Nebraska to Texas east to the Atlantic, in extreme northern Ohio, southern Ontario, and southern California. *Solanum dulcamara* is a naturalized weed from Eurasia that is now common in the northern United States and Canada, and south in the mountains of North Carolina and Tennessee. *Solanum linnaeanum* is a common weed in Hawaii. *Solanum mammosum* grows in the West Indies and tropical America. *Solanum pseudocapsicum*, a decorative pot plant, has escaped from cultivation in Hawaii and the Gulf Coast states. *Solanum seaforthianum* is a South American plant cultivated in warmer areas, including Hawaii; it has become naturalized in Florida. *Solanum tuberosum*, the white potato of commerce, is a widely cultivated vegetable. An occasional plant escapes from cultivation or from dumps.

Toxic Part: In *Solanum tuberosum*, the uncooked sprout and sun-greened skin are toxic. In the remaining species, human poisoning is generally attributed to immature fruit. Several species produce dermatitis.

Solanum tuberosum

Toxin: Solanine glycoalkaloids have predominantly gastrointestinal irritant effects. There have been reports of atropine-like poisoning, but atropine and related alkaloids are not generally found in these plants.

Clinical Findings: Nausea, vomiting, abdominal cramping, and diarrhea may occur. Central nervous system effects of delerium, hallucinations, and coma have been reported, but the mechanisms for these effects are not known.

Management: Intravenous hydration, antiemetics, and electrolyte replacement may be necessary for patients with severe gastrointestinal effects, particularly in children. Central nervous system effects are managed with supportive measures and typically resolve without sequalae. Consultation with a Poison Control Center should be considered. See "Poisoning by Plants with Gastrointestinal Toxins," p. 28.

References

Ceha LJ, Presperin C, Young E, Allswede M, Erickson T. Anticholinergic toxicity from nightshade berry poisoning responsive to physostigmine. J Emerg Med 1997;15(1): 65–69.

Hornfeldt CS, Collins JE. Toxicity of nightshade berries (Solanum dulcamara) in mice. J Toxicol Clin Toxicol 1990;28:185–192.

Korpan YI, Nazarenko EA, Skryshevskaya IV, Martelet C, Jaffrezic-Renault N, El'skaya AV. Potato glycoalkaloids: True safety or false sense of security? Trends Biotechnol 2004;22: 147–151.

McMillan M, Tompson JC. An outbreak of suspected solanine poisoning in schoolboys. Examination of solanine poisoning. Q J Med 1979;48:227–243.

Patil BC, Sharma RP. Evaluation of solanine toxicity. Food Cosmet Toxicol 1972;10: 395–398.

Phillips BJ, Hughes JA, Phillips JC, Walters DG, Anderson D, Tahourdin CS. A study of the toxic hazard that might be associated with the consumption of green potato tops. Food Chem Toxicol 1996;34:439–448.

Sophora species
Family: Leguminosae (Fabaceae)
Sophora secundiflora
  (Ortega) Lag. ex DC.
Sophora tomentosa L.

Common Names:
Sophora secundiflora: Burn Bean, Coral Bean, Colorines, Frijolillo, **Mescal Bean**, Pagoda Tree, Red Bean, Red Hots, Texas Mountain Laurel
Sophora tomentosa: **Necklacepod Sophora**, Silver Bush, Tambalisa

Sophora secundiflora, fruiting branch (above)

Sophora secundiflora, close-up of flowers (below)

Description: *Sophora secundiflora* is a shrub or small tree with small compound leaflets; the leaflets grow to 2 inches in length and occur in three to five pairs. The purple flowers cluster in showy racemes. Seeds are bright red and are contained in woody fruit pods. *Sophora tomentosa* is a shrub with numerous branches and compound leaves. Bright yellow flowers grow in racemes. The yellow seeds are pealike and are contained in fruit pods.

Distribution: *Sophora secundiflora* grows in Texas, New Mexico, Mexico, the coastal dunes of Baja, California, Hawaii, and Guam. Its brightly colored seeds are used in jewelry. *Sophora tomentosa* grows in Florida, Bermuda, and the West Indies from the Bahamas to Barbados.

Toxic Part: The seeds are most poisonous but have a hard coat. The leaves and flowers contain the toxin as well.

Toxin: Cytisine and related nicotine-like alkaloids.

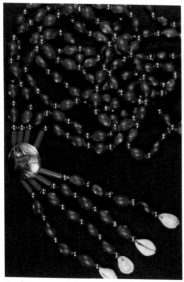

Sophora secundiflora, necklace made from seeds

Sophora secundiflora, fruit with mature seeds

Clinical Findings: There are no adequately documented human poisonings, and clinical descriptions are derived primarily from animal reports. Initial gastrointestinal effects may be followed by those typical of nicotine poisoning; these include hypertension, large pupils, sweating, and perhaps seizures. Severe poisoning produces coma, weakness, and paralysis that may result in death from respiratory failure. Despite the similar-sounding names, mescal bean does not contain mescaline.

Management: Symptomatic and supportive care should be given, with attention to adequacy of ventilation and vital signs. Atropine may reverse some of the toxic effects. Consultation with a Poison Control center should be strongly considered. See "Poisoning by Plants with Nicotine-like Alkaloids," p. 30.

References

Bourn WM, Keller WJ, Bonfiglio J. Comparisons of mescal bean alkaloids with mescaline, delta 9-THC and other psychotogens. Life Sci 1979;25:1043–1054.

Izaddoost M, Harris BG, Gracy RW. Structure and toxicity of alkaloids and amino acids of *Sophora secundiflora*. J Pharm Sci 1976;65:352–354.

Knauer KW, Reagor JC, Bailey EM Jr, Carriker L. Mescalbean (*Sophora secundiflora*) toxicity in a dog. Vet Hum Toxicol 1995;37:237–239.

Sullivan G, Chavez PI. Mexican good-luck charm potentially dangerous. Vet Hum Toxicol 1980;23:259–260.

Wellmann KF. North American Indian rock art and hallucinogenic drugs. JAMA 1978;239:1524–1527.

Spathiphyllum species
Family: Araceae

Common Names: Peace Lily, Spathe Flower, White Anthurium

Description: *Spathiphyllum* species are tropical perennial herbs that grow to about 2 feet. The elliptical leaves grow on short stems and emerge in clusters. The flower has a white or greenish spathe with a short, usually white, spadix resembling a miniature ear of corn.

Distribution: These plants are primarily of South American origin and may be grown outdoors in subtropical climates. They are commonly cultivated as indoor plants elsewhere.

Toxic Part: All parts of this plant are injurious.

Toxin: Raphides of water-insoluble calcium oxalate and unverified proteinaceous toxins.

Clinical Findings: A painful burning sensation of the lips and mouth results from ingestion. There is an inflammatory reaction, often with edema and blistering. Hoarseness, dysphonia, and dysphagia may result.

Management: The pain and edema recede slowly without therapy. Cool liquids or demulcents held in the mouth may bring some relief. Analgesics may be

Spathiphyllum sp.

Spathiphyllum sp., close-up of flowers showing spadix enclosed by spathe

indicated. The insoluble oxalate in these plants does not cause systemic oxalate poisoning. Consultation with a Poison Control Center should be considered. See "Poisoning by Plants with Calcium Oxalate Crystals," p. 23.

References

Franceschi VR, Nakata PA. Calcium oxalate in plants: formation and function. Annu Rev Plant Biol 2005;56:41–71.

Kanerva L, Estlander T, Aalto-Korte K. Occupational protein contact dermatitis and rhinoconjunctivitis caused by spathe (*Spathiphyllum*) flowers. Contact Dermatitis 2000;42:369–370.

Krenzelok EP, Mrvos R, Jacobsen TD. Contrary to the literature, vomiting is not a common manifestation associated with plant exposures. Vet Hum Toxicol 2002;44:298–300.

Spigelia species
Family: Loganiaceae
Spigelia anthelmia L.
Spigelia marilandica L.

Common Names:
Spigelia anthelmia: Espigelia, Herbe-à-Brinvilliers, Loggerhead Weed, Lombricera, Pink Weed, Waterweed, **West Indian Pinkroot**, Worm Grass
Spigelia marilandica: Carolina Pink, Indian Pink, **Pinkroot**, Worm Grass

Description: *Spigelia anthelmia* is an annual herb that grows to about 1.5 feet. The opposite leaves have very short stems and are about 6 inches long and 3 inches wide. The tubular flowers grow in a terminal spike and flare into five petals. Petals are pink; the tube is white with magenta stripes. Fruits are globose and two lobed. *Spigelia marilandica* is a perennial herb growing to 2 feet. The ovate opposite leaves are about 4 inches long. The tubular flowers are about 2 inches long, red on the outside, and yellow on the inside.

Distribution: *Spigelia anthelmia* grows in southern Florida, the Bahamas, and the West Indies to Barbados. *Spigelia marilandica* grows in Florida to Texas north to southern Indiana and North Carolina.

Toxic Part: The whole plant is poisonous.

Toxin: Spigeline, a strychnine-like alkaloid.

Clinical Findings: Toxicity from this plant is not commonly reported. In instances of toxicity, gastrointestinal effects occur early and are followed by muscle spasms that can be severe and strychnine like. If extreme muscle activity occurs, then rhabdomyolysis and renal injury are concerns.

Management: Muscle spasms should be controlled with benzodizepines. Fluids and electrolyte balance needs to be maintained to avoid dehydration and minimize the renal complications of rhabdomyolysis. Consultation with a

Spigelia anthelmia

Spigelia marilandica, close-up of flower

Poison Control Center should be strongly considered. See "Poisoning by Plants with Convulsive Poisons (Seizures)," p. 25.

References

Achenbach H, Hubner H, Vierling W, et al. Spiganthine, the cardioactive principle of *Spigelia anthelmia*. J Nat Prod 1995;58:1092–1096.

Camurca-Vasconcelos AL, Nascimento NR, et al. Neuromuscular effects and acute toxicity of an ethyl acetate extract of *Spigelia anthelmia* Linn. J Ethnopharmacol 2004;92:257–261.

Esposito-Avella M, Brown P, Tejeira I, et al. Pharmacological screening of Panamanian medicinal plants. Part 1. J Crude Drug Res 1985;23:17–25.

Strychnos nux-vomica L.
Family: Loganiaceae

Common Names: Nux-Vomica Tree, **Strychnine**

Description: This small tree has ovate leaves growing to about 2 by 3.5 inches. The yellowish-white tubular flowers grow in terminal clusters. The fruit is 1.5 inches across and hard shelled; it varies in color from yellow to orange and resembles a small grapefruit. It contains several gray velvety seeds that look like nickel-sized buttons.

Distribution: *Strychnos nux-vomica* is native to Asia, and cultivated in Hawaii; it is used to produce strychnine.

Strychnos nux-vomica, fruiting branch

Toxic Part: The whole plant, including the seeds, is poisonous.

Toxin: Strychnine.

Clinical Findings: Poisoning from plant ingestion is rare. Despite the known toxicity, the seeds continue to be used in some countries as a medicinal herbal product. Initially the patient may experience feelings of restlessness and anxiety, followed by fasciculations and hyperreflexia. In severe poisoning, generalized symmetrical tonic spasms with opisthotonic posturing develop. Although resembling seizures, consciousness is maintained. The extreme muscle spasms can last from seconds to minutes and are triggered by movement or excitement. Rhabdomyolysis and hyperthermia can develop.

Management: Aggressive control of muscle spasms is the goal of therapy. Use of intravenous benzodiazepines may not be adequate. Persistent convulsions may require use of a neuromuscular blockade. Hyperthermia may need to be managed with external cooling measures. Fluid and electrolyte balance needs to be maintained to avoid dehydration and minimize the renal complications of rhabdomyolysis. Consultation with a Poison Control Center should be strongly considered. See "Poisoning by Plants with Convulsive Poisons (Seizures)," p. 25.

References

Philippe G, Angenot L, Tits M, Frederich M. About the toxicity of Strychnos species and their alkaloids. Toxicon 2004;44:405–416.

Ryan CJ, Anderson J. Strychnine poisoning. N Engl J Med 2001;345:1577.

Starretz-Hacham O, Sofer S, Lifshitz M. Strychnine intoxication in a child. Isr Med Assoc J 2003;5:531–532.

Wood D, Webster E, Martinez D, Dargan P, Jones A. Survival after deliberate strychnine self-poisoning, with toxicokinetic data. Crit Care 2002;6:456–459.

Zhang YG, Huang GZ. Poisoning by toxic plants in China: report of 19 autopsy cases. Am J Forensic Med Pathol 1988;9:313–319.

Swietenia mahagoni (L.) Jacq.
Family: Meliaceae

Common Names: Caoba, **Mahogany**

Swietenia mahagoni (above)

Swietenia mahagoni, close-up of fruit and leaves (left)

Description: This large tree has dark brown, rough bark and alternate, pinnately compound leaves with an even number of leaflets. The tree is semideciduous; older leaves are discarded in the spring during the period of new growth. The flowers occur in inconspicuous clusters. Usually a single flower in each cluster develops into an erect, five-lobed, woody fruit pod, which splits from the base to release winged seeds about 1 inch long.

Distribution: The mahogany is native to south Florida, the Bahamas, Cuba, Jamaica, Haiti, the Dominican Republic, and Puerto Rico. It is cultivated in Hawaii.

Toxic Part: Seeds are poisonous.

Toxin: An unidentified toxin.

Clinical Findings: A single case report involving an adult describes gastrointestinal, central nervous system, and cardiac effects after the ingestion of two seeds. The patient had multiple episodes of vomiting followed by coma, bradycardia, and hypotension. He was treated with fluids, atropine, and vasopressors, with recovery occurring over the ensuing 48 hours.

Management: If poisoning occurs, then treatment is supportive in nature. Intravenous hydration, antiemetics, and electrolyte replacement may be necessary for patients with severe gastrointestinal effects. Cardiovascular support may be necessary if bradycardia or hypotension develop. Consultation with a Poison Control Center should be considered.

Reference

Raghuraman V, Raveendran M. Mahogany seeds: An unusual poison. J Indian Med Assoc 1982;78:186–188.

Symphoricarpos species
Family: Caprifoliaceae
Symphoricarpos albus (L.) S.F. Blake (=*S. racemosus* Michx.)
Symphoricarpos occidentalis Hook.
Symphoricarpos orbiculatus Moench

Common Names:
Symphoricarpos albus: Belluaine, Buck Brush, **Snowberry**, Waxberry
Symphoricarpos occidentalis: Western Snowberry, **Wolfberry**
Symphoricarpos orbiculatus: Coralberry, Devil's Shoestring, **Indian Currant**

Description: These deciduous shrubs grow to about 3 to 7 feet. Leaves are simple, opposite, up to 3 inches long. The small bell-shaped flowers of *Symphoricarpos albus* and *S. occidentalis* are pink, and those of *S. orbiculatus* are white. They grow in small clusters along the main stem. The fruit of *Symphoricarpos albus* and *S. occidentalis* is a white berry, while that of *S. orbiculatus* is coral red. The fruits contain two seeds.

Distribution: Snowberry grows in woods and open slopes in southeastern Alaska (but not in the Yukon), Alberta to Quebec, south to California, Colorado, Nebraska, and eastern North America. Indian currant grows in eastern North America and Texas. Wolfberry is most common in the northern Great Plains, and grows from British Columbia to Ontario, as well as Washington, Utah, New Mexico, Oaklahoma, Missouri, and Michigan. *Symphoricarpos* species are cultivated elsewhere as well.

Symphoricarpos albus, close-up of fruit

Toxic Part: The berries are toxic.

Toxin: Saponin, and small quantities of chelidonine, an isoquinoline alkaloid. (See also *Chelidonium majus*, p. 120.)

Clinical Findings: With substantial ingestion of berries, nausea, vomiting, abdominal cramping, and diarrhea may occur.

Management: Intravenous hydration, antiemetics, and electrolyte replacement may be necessary for patients with severe gastrointestinal effects, particularly in children. Consultation with a Poison Control Center should be considered. See "Poisoning by Plants with Gastrointestinal Toxins," p. 28.

Symphoricarpos orbiculatus, close-up of fruit and leaves

References

Lamminpaa A, Kinos M. Plant poisonings in children. Hum Exp Toxicol 1996;15(3):245–249.

Lewis WH. Snowberry (*Symphoricarpos*) poisoning in children. JAMA 1979;242:2663.

Symplocarpus foetidus (L.) Nutt.
Family: Araceae

Common Names: Chou Puant, Polecat Weed, **Skunk Cabbage**, Tabac du Diable

Description: The flowering spathe appears before the leaves in the spring. The spathe is 3 to 6 inches long and may be green, purple, or brown, and striped or spotted. The leaves eventually reach 3 feet in length and 1 foot in width. All parts of this plant have an unpleasant odor.

Distribution: Skunk cabbage grows from Quebec to Nova Scotia to North Carolina and Iowa. Its natural habitats are moist areas such as wet woods and swamps.

Toxic Part: The leaves are injurious.

Toxin: Raphides of water-insoluble calcium oxalate and unverified proteinaceous toxins.

Clinical Findings: A painful burning sensation of the lips and mouth results from ingestion. There is an inflammatory reaction, often with edema and blistering. Hoarseness, dysphonia, and dysphagia may result.

Symplocarpus foetidus, leaves

Symplocarpus foetidus, close-up of flowers

Management: The pain and edema recede slowly without therapy. Cool liquids or demulcents held in the mouth may bring some relief. Analgesics may be indicated. The insoluble oxalate in these plants does not cause systemic oxalate poisoning. Consultation with a Poison Control Center should be considered. See "Poisoning by Plants with Calcium Oxalate Crystals," p. 23.

References

Rauber A. Observations on the idioblasts of *Dieffenbachia*. J Toxicol Clin Toxicol 1985;23: 79–90.

Watson JT, Jones RC, Siston AM, et al. Outbreak of food-borne illness associated with plant material containing raphides. Clin Toxicol (Phila) 2005;43:17–21.

Taxus species
Family: Taxaceae
Taxus baccata L.
Taxus brevifolia Nutt.
Taxus canadensis Marsh.
Taxus cuspidata Siebold & Zucc.
Taxus floridana Nutt.

Common Names: Ground Hemlock, **Yew**
Taxus baccata: **English Yew**
Taxus brevifolia: **Western Yew**
Taxus canadensis: **Canadian Yew**, Buis de Sapia
Taxus cuspidata: **Japanese Yew**
Taxus floridana: **Florida Yew**

Description: These *Taxus* species are evergreen trees and shrubs with alternate branchlets and reddish-brown, thin, scaled bark. They can grow to 60 feet or

Taxus brevifolia

Taxus brevifolia, branch with fruit

more, but some species are commonly pruned into hedges or low shrubs. The flat, alternate, needle-like leaves are about 1 inch long. The hard seeds are green to black and partially exposed in a fleshy red cup (aril).

Taxus cuspidata cv. 'Aurescens,' shrub

Distribution: *Taxus baccata* is cultivated in the southern United States. *Taxus brevifolia* grows from Alaska south along coastal British Columbia to western Washington, Oregon, northern California, Idaho, and Montana and southern Alberta. *Taxus canadensis* grows in North Carolina, Virginia, Pennsylvania, West Virginia to Iowa north to southeastern Manitoba to Nova Scotia. *Taxus cuspidata* is cultivated in the north temperate zones. *Taxus floridana* is found in northwestern Florida.

Toxic Part: Most of the plant including the seeds, but *not* the red aril, is toxic.

Toxin: Taxine alkaloids, block sodium and calcium currents.

Clinical Findings: Most pediatric cases involve ingestions of the seed and aril, and patients are minimally symptomatic. Substantial ingestion of the leaves causes gastrointestinal, neurological, and cardiovascular effects. Dizziness, dry mouth, and mydriasis develop within 1 hour, followed by abdominal cramping, salivation, and emesis. There is generalized weakness, and the patient may become comatose. Sinus bradycardia, premature ventricular contractions, atrioventricular conduction defects, or ventricular tachydysrhythmias may occur. Hyperkalemia, if present, may be an indicator of toxicity. Deaths are caused by cardiac dysrhythmias. Allergic reactions may result from chewing the needles.

Management: Gastrointestinal decontamination as appropriate, serial electrocardiograms, and serum potassium determinations should be performed. Because of the clinical similarity to cardioactive steroid toxicity, digoxin-specific Fab fragments have been used to treat serious dysrhythmias, but there is no evidence that cardioactive steroids are present in Yew. Consultation with a Poison Control Center should be strongly considered.

References

Cummins RO, Haulman J, Quan L. Near-fatal yew berry intoxication treated with external cardiac pacing and digoxin-specific Fab antibody fragments. Ann Emerg Med 1990;19:77–82.

Ingen GV, Visser R, Peltenbury H, et al. Sudden unexpected death due to *Taxus* poisoning. A report of five cases with review of the literature. Forensic Sci Int 1992;56:81–87.

Ruha AM, Tanen DA, Graeme KA, et al. Hypertonic sodium bicarbonate for *Taxus media*-induced cardiac toxicity in swine. Acad Emerg Med 2002;9:179–185.

Werth J, Murphy JJ. Cardiovascular toxicity associated with yew leaf ingestion. Br Heart J 1994;72:92–93.

Willaert W, Claessens P, Vankelecom B, Vanderheyden M. Intoxication with *Taxus baccata*: cardiac dysrhythmias following yew leaves ingestion. Pacing Clin Electrophysiol 2002;5:511–512.

Wilson CR, Sauer J, Hooser SB. Taxines: A review of the mechanism and toxicity of yew (*Taxus* spp.) alkaloids. Toxicon 2001;39:175–185.

Thevetia peruviana (Pers.) K. Schum.
(=*T. neriifolia* A. Juss. ex Steud. =*Cascabela thevetia* (L.) Lippold)
Family: Apocynaceae

Common Names: Ahouai des Antilles, Be-Still Tree, Cablonga, Flor del Perú, Lucky Nut, Noho-Malie, Retama, Serpent, **Yellow Oleander**

Description: This small tree is 10 to 20 feet tall with leaves resembling those of the *Nerium oleander*, but *Thevetia* has a milky sap. Flowers are yellow with a peach tinge. The small (about 1 inch), clam-shaped fruit contains two to four flat seeds.

Distribution: *Thevetia* grows in south Florida and the southwestern United States, the West Indies, Sri Lanka, Hawaii, and Guam.

Toxic Part: All parts of the plant, particularly the seeds, are poisonous.

Toxin: Thevetin A & B, thevetoxin, and peruvoside, cardioactive steroids resembling digitalis.

Clinical Findings: Poisoning produces clinical findings typical of cardioactive steroid poisoning. Toxicity has a variable latent period that depends on the quantity ingested. Dysrhythmias are usually expressed as sinus bradycardia, premature ventricular contractions, atrioventricular conduction defects, or ventricular tachydysrhythmias. Hyperkalemia, if present, may be an indicator of toxicity.

Thevetia peruviana, branch with flower

Management: Gastrointestinal decontamination as appropriate, serial electrocardiograms, and serum potassium determinations should be performed. If serious cardioactive steroid toxicity is considered, digoxin-specific Fab should be administered. Consultation with a Poison Control Center should be strongly considered. See "Poisoning by Plants with Cardioactive Steroids/Cardiac Glycosides," p. 24.

References

De Silva HA, Fonseka MM, Pathmeswaran A, et al. Multiple dose activated charcoal for treatment of yellow oleander poisoning: A single-blind, randomised, placebo-controlled trial. Lancet 2003;361:1935–1938.

Eddleston M, Ariaratnam CA, Sjostrom L, et al. Acute yellow oleander (*Thevetia peruviana*) poisoning: cardiac dysrhythmias, electrolyte disturbances, and serum cardiac glycoside concentrations on presentation to hospital. Heart 2000;83:301–306.

Eddleston M, Rajapakse S, Rajakanthan, et al. Anti-digoxin Fab fragments in cardiotoxicity induced by ingestion of yellow oleander: A randomised controlled trial. Lancet 2000;355: 967–972.

Eddleston M, Ariaratnam CA, Meyer WP, et al. Epidemic of self-poisoning with seeds of the yellow oleander tree (*Thevetia peruviana*) in northern Sri Lanka. Trop Med Int Health 1999;4:266–273.

Toxicodendron species
Family: Anacardiaceae

Toxicodendron diversilobum (Torr. & A. Gray) Greene (=*Rhus diversiloba* Torr. & A. Gray)

Toxicodendron pubescens P. Mill. (=*Rhus toxicarium* Salisb.; *Toxicodendron toxicarium* (Salisb.) Gillis)

Toxicodendron radicans (L.) Kuntze (=*Rhus radicans* L.)

Toxicodendron rydbergii (Small) Greene

Toxicodendron vernix (L.) Kuntze (=*Rhus vernix* L.)

Toxicodendron diversilobum

Common Names:
Toxicodendron diversilobum: **Western Poison Oak**
Toxicodendron pubescens: **Eastern Poison Oak**
Toxicodendron radicans: **Poison Ivy**
Toxicodendron rydbergii: **Western Poison Ivy**
Toxicodendron vernix: **Poison Sumac**

Description: *Toxicodendron diversilobum* is an erect shrub 3 to 6 feet tall with three round or even-lobed leaves and producing white fruits. *Toxicodendron pubescens* is a simple or sparingly branched shrub growing to about 3 feet, with 3 blunt-lobed leaflets and yellowish-white fruits. *Toxicodendron radicans* is a climbing vine covered with numerous aerial roots, with 3 ovate to elliptical leaflets with pointed teeth or shallow lobes. *Toxicodendron ryd-*

Toxicodendron radicans

Toxicodendron radicans, with Autumn color

bergii is a shrub growing 3 to 9 feet tall, with leaves having pointed teeth on their margins. *Toxicodendron vernix* is a shrub to small tree, 6 to 15 feet tall, branched at the base, with compound leaves having 7 to 13 leaflets with smooth edges and producing grayish-white fruits.

Distribution: Western Poison Oak (West of the Rocky Mountains): Eastern Poison Oak (New Jersey southward); Poison Ivy (Eastern North America); Western Poison Ivy (widespread in the Western United States extending East into the Northern United States and Southern Canada to Nova Scotia and New

Toxicodendron vernix

York); Poison Sumac (swampy areas in Florida, Northeast and Nova Scotia to Minnesota).

Toxic Part: All parts of the plant contain the oily resin that contains the toxin.

Toxin: Urushiol, an antigenic catechol.

Clinical Findings: All these plants can cause mild to severe contact dermatitis in patients previously sensitized to urushiols, and there is cross-reactivity among the various species. The skin rash typically develops within 24 hours following exposures, although in more highly allergic individuals it may develop within several hours. The rash often appears in a linear pattern consistent with brushing up against the plant. More than 50% of the population is sensitized to urushiol and will get rhus dermatitis on contact with a plant of this genus.

Management: Symptomatic and supportive care; this includes the potential use of topical or oral antihistamines and possibly corticosteroids. Consultation with a Poison Control Center should be considered.

References

Crawford GH, McGovern TW. Poison ivy. N Engl J Med 2002;347:1723–1724.

Guin JD. Treatment of toxicodendron dermatitis (poison ivy and poison oak). Skin Ther Lett 2001;6:3–5.

Oh SH, Haw CR, Lee MH. Clinical and immunologic features of systemic contact dermatitis from ingestion of *Rhus* (*Toxicodendron*). Contact Dermatitis 2003;48:251–254.

Park SD, Lee SW, Chun JH, Cha SH. Clinical Future if 31 patients with systemic contact dermatitis due to the ingestion of Rhus (lagver). Br J Dermatol 2000;142:937–942.

Urginea maritima **(L.) Baker**
(=*Scilla maritima* L.)
Family: Liliaceae

Common Names: Red Squill, Sea Onion, Squill

Description: *Urginea* grows from an onion-like bulb. The leaves are 1.5 feet long and 4 inches wide. Whitish to pink flowers appear in dense clusters at the tips of the stems.

Distribution: Red squill is native to Eurasia, the Mediterranean, and South Africa. It is primarily in cultivation for the commercial extraction of squill (known also as red squill) for use as a rat poison.

Toxic Part: The bulbs are poisonous.

Toxin: Scillarin, a cardio-active steroid resembling digitalis.

Clinical Findings: Substantial ingestion may lead to toxicity. Poisoning produces clinical findings typical of cardioactive steroid poisoning. Toxicity has a variable latent period that depends on the quantity ingested. Dysrhythmias are usually expressed as sinus bradycardia, premature ventricular contractions, atrioventricular conduction defects, or ventricular tachydysrhythmias. Hyperkalemia, if present, may be an indicator of toxicity.

Management: Gastrointestinal decontamination as appropriate, serial electrocardiograms, and serum potassium determinations should be performed. If serious cardioactive steroid toxicity is considered, dig-oxin-specific Fab should be administered. Consultation with a Poison Control Center should be considered. See "Poisoning by Plants with Cardioactive Steroids/Cardiac Glycosides," p. 24.

Urginea maritima

References

el Bahri L, Djegham M, Makhlouf M. *Urginea maritima* L. (Squill): A poisonous plant of North Africa. Vet Hum Toxicol 2000;42:108–110.

Foukaridis GN, Osuch E, Mathibe L, Tsipa P. The ethnopharmacology and toxicology of *Urginea sanguinea* in the Pretoria area. J Ethnopharmacol 1995;49:77–79.

Tuncok Y, Kozan O, Cavdar C, Guven H, Fowler J. *Urginea maritima* (squill) toxicity. J Toxicol Clin Toxicol 1995;33:83–86.

Veratrum species
Family: Liliaceae
Veratrum album L.
Veratrum californicum Durand
Veratrum parvifolium Michx.
Veratrum tenuipetalum A. Heller
Veratrum viride Aiton

Common Names: American White Hellebore, Corn Lily, Earth Gall, Green Hellebore, **False Hellebore**, Indian Poke, Itch Weed, Pepper-Root, Rattlesnake Weed, Skunk Cabbage, Swamp Hellebore, Tickle Weed, White Hellebore

Description: *Veratrum* species are tall perennial herbs with alternate, pleated leaves. The flowers are white, marked with green on the top portion of the stalk. The fruit is a small pod containing winged seeds.

Distribution: *Veratrum album* grows in the Aleutian Islands, Alaska. *Veratrum californicum* grows on the West Coast from Washington to Baja, California, east to Montana, Colorado, and New Mexico. *Veratrum parvifolium* grows in the southern Appalachian Mountains. *Veratrum tenuipetalum* grows in Colorado and Wyoming. *Veratrum viride* grows in Alaska, the Yukon, British Columbia, Alberta, Oregon, Montana, Minnesota, and Quebec south to Tennessee and Georgia.

Veratrum californicum

Toxic Part: All parts of this plant are poisonous.

Toxin: Veratrum alkaloids, sodium channel activators.

Clinical Findings: Symptoms are predominantly neurological and cardiac. There is transient burning in the mouth after ingestion, followed after several hours by increased salivation, vomiting, diarrhea, and a prickling sensation in the skin. The patient may complain of headache, muscular weakness, and dimness of vision. Bradycardia and other cardiac dysrhythmias can be associated with severe blood pressure abnormalities. Coma may develop, and convulsions may be a terminal event.

Management: Fluid replacement should be instituted with respiratory support if indicated. Heart rhythm and blood pressure should be monitored and treated with appropriate medications and supportive care. Recovery is generally complete within 24 hours. Consultation with a Poison Control Center should be strongly considered. See "Poisoning by Plants with Sodium Channel Activators," p. 32.

Veratrum viride

References

Dunnigan D, Adelman RD, Beyda DH. A young child with altered mental status. Clin Pediatr (Phila) 2002;41:43–45.

Gaillard Y, Pepin G. LC-EI-MS determination of veratridine and cevadine in two fatal cases of *Veratrum viride* poisoning. J Anal Toxicol 2001;25:481–485.

Jaffe AM, Gephardt D, Courtemanche L. Poisoning due to ingestion of *Veratrum viride* (false hellebore). J Emerg Med 1990;8:161–167.

Prince LA, Stork CM. Prolonged cardiotoxicity from poisonlilly (*Veratrum viride*). Vet Hum Toxicol 2000;42:282–285.

Quatrehomme G, Bertrand F, Chauvet C, et al. Intoxication from *Veratrum album*. Hum Exp Toxicol 1993;12:111–115.

Viscum album L.
Family: Loranthaceae (Viscaceae)

Viscum album, branch with immature fruit

Common Names: European Mistletoe, Mistletoe

Description: This parasite grows primarily on the trunks of deciduous trees, particularly the apple. Stems are much branched, and the leaves are 2 to 3 inches long, thick, leathery, and usually a pale yellowish-green. The fruit is a sticky white berry.

Distribution: This European plant was introduced into Sonoma County, California.

Toxic Part: Only the leaves and stems are toxic. The berries have very low toxin concentration and cause toxicity only if ingested in large quantities.

Toxin: Viscumin, a plant lectin (toxalbumin) related to ricin.

Clinical Findings: The toxin is similar in action to the lectins contained in *Abrus precatorius* and *Ricinus communis* but is less potent. Following a latent period of many hours, nausea, vomiting, abdominal cramping, diarrhea, and dehydration occur.

Management: Patients with symptoms need to be assessed for signs of dehydration and electrolyte abnormalities. Activated charcoal should be administered. Intravenous hydration, antiemetics, and electrolyte replacement may be necessary in severe cases, particularly those involving children. Consultation with a Poison Control Center should be considered. See "Poisoning by Plants with Toxalbumins," p. 33.

References

Franz H. Mistletoe lectins and their A and B chains. Oncology 1986;43(suppl 1): 23–34.

Gorter RW, van Wely M, Stoss MM, Wollina U. Subcutaneous infiltrates induced by injection of mistletoe extracts (Iscador). Am J Ther 1998;5:181–187.

Stirpe F. Mistletoe toxicity. Lancet 1983;1:295.

Wisteria species
Family: Leguminosae (Fabaceae)
Wisteria floribunda (Willd.) DC.
Wisteria sinensis (Sims) Sweet

Common Names: Kidney Bean Tree, **Wisteria**, Wistaria

Description: These woody vines bear drooping masses of sweetpea-like flowers, which are usually blue, but pink and white varieties also exist. The fruit pods of *Wisteria floribunda* are smooth; those of *W. sinensis* are covered with velvety down. The pods persist through winter.

Distribution: Wisterias are hardy in the north but are most common in the southeastern United States as far west as Texas.

Wisteria floribunda, flowering vines

Toxic Part: All parts of this plant are toxic. Although the flowers are sometimes considered nontoxic, reports suggest otherwise.

Toxin: Wistarine, a glycoside.

Clinical Findings: Nausea, vomiting, abdominal cramping, and diarrhea may occur.

Management: Intravenous hydration, antiemetics, and electrolyte replacement may be necessary for patients with severe gastrointestinal effects, particularly in children. Consultation with a Poison Control Center should be considered. See "Poisoning by Plants with Gastrointestinal Toxins," p. 28.

Wisteria sinensis, leaves with flowers

Wisteria sinensis, fruit with seeds

References

Piola C, Ravaglia M, Zoli MP. Poisoning of *Wisteria sinensis* seeds. Two clinical cases. Boll Soc Ital Farm Osp 1983;29:333–337

Rondeau ES. Wisteria toxicity. J Toxicol Clin Toxicol 1993;31:107–112.

Xanthosoma species
Family: Araceae

Common Names: 'Ape, Blue 'Ape, Blue Taro, Caraibe, **Malanga**, Yautía

Description: *Xanthosoma* species resemble *Caladium*, but the leaves are more spear shaped. The tubers or rhizomes are thick, and the sap is milky.

Distribution: This plant is in cultivation in the southern United States, Hawaii, the West Indies, and Guam.

Toxic Part: The leaves are injurious. Some species are grown for their edible tubers.

Toxin: Raphides of water-insoluble calcium oxalate and unverified proteinaceous toxins.

Xanthosoma violacea

Xanthosoma sagittifolium, close-up of flower

Clinical Findings: A painful burning sensation of the lips and mouth may result from ingestion. There is an inflammatory reaction, often with edema and blistering. Hoarseness, dysphonia, and dysphagia may result.

Management: The pain and edema recede slowly without therapy. Cool liquids or demulcents held in the mouth may bring some relief. Analgesics may be indicated. The insoluble oxalate in these plants does not cause systemic oxalate poisoning. Consultation with a Poison Control Center should be considered. See "Poisoning by Plants with Calcium Oxalate Crystals," p. 23.

Reference

Sakai WS, Hanson M, Jones RC. Raphides with barbs and grooves in *Xanthosoma sagittifolium* (Araceae). Science 1972;178:314–315.

Zamia species
Family: Zamiaceae
Zamia furfuracea Aiton
Zamia integrifolia Aiton
Zamia pumila L.

Common Names: Bay Rush, Camptie, Cardboard Palm, **Coontie**, Guayiga, Florida Arrowroot, Marunguey, Mexican Cycad, Palmita de Jardín, Sago Cycas, Seminole Bread, Yugulla

Description: The short trunk of these *Zamia* species may be entirely underground; *Z. furfuracea* has 6 to 30 or rarely 40 leaves, with sparsely to densely prickled petioles, whereas *Z. integrifolia* has 2 to 15 leaves with smooth petioles. *Zamia pumila* has 2 to 15 leaves and petioles with stipules. The leaves are pinnate, somewhat palmlike, and both male and female cones are produced.

Distribution: *Zamia integrifolia* grows on the southeastern coast of Georgia and throughout Florida, as well as on the Bahamian Islands, Cuba, and the Cayman Islands. Found at one time in Puerto Rico, it is now probably extinct there. *Zamia furfuracea* is native to southeastern Veracruz, Mexico, and is cultivated elsewhere. *Zamia pumila* is native to the Dominican Republic and Central Cuba.

Toxic Part: All parts of all *Zamia* species are poisonous. The toxin can be removed from the grated root by water; treated plant material was a commercial source of starch in Florida until the 1920s.

Toxin: Cycasin, an azoglycoside that releases methylazoxymethanol (responsible for toxicity) on hydrolysis.

Clinical Findings: If poisoning occurs, the most common effects are nausea, vomiting, and abdominal cramping. Other symptoms may include visual complaints, lethargy, and, in extreme cases, coma.

Zamia integrifolia

Zamia pumila

Zamia pumila, with cone

Management: Intravenous hydration, antiemetics, and electrolyte replacement may be necessary for patients with severe gastrointestinal symptoms, particularly in children. There are no specific antidotes for the CNS manifestations. Consultation with a Poison Control Center should be considered. See "Poisoning by Plants with Gastrointestinal Toxins," page 28.

References

Hall WTK. Cycad (*Zamia*) poisoning in Australia. Aust Vet J 1987;64:149–151.

Norstog KJ, Nichols TJ. *The Biology of the Cycads.* Cornell University Press, Ithaca, 1997.

Schneider D, Wink M, Sporer F, Lounibos P. Cycads: Their evolution, toxins, herbivores and insect pollinators. Naturwissenschaften 2002;89:281–294.

Yagi F. Azoxyglycoside content and beta-glycosidase activities in leaves of various cycads. Phytochemistry 2004;65:3243–3247.

Zantedeschia aethiopica (L.) Spreng.
Family: Araceae

Common Names: Arum Lily, Calla, **Calla Lily**, Lirio Cala

Description: This is the calla of gardeners or the "flowers of florists," not the true calla (*Calla palustris*). This plant has smooth-edged, arrowhead-shaped leaves that are sometimes mottled with white and grow on long, stout stalks. The showy flowering spathe flares out like a lily. It is white or green in this species but may be pink or yellow in others. The spadix is yellow.

Distribution: *Zantedeschia* is grown outdoors in mild climates and as a greenhouse or houseplant elsewhere.

Toxic Part: The leaves are injurious.

Toxin: Raphides of water-insoluble calcium oxalate and unverified proteinaceous toxins.

Zantedeschia aethiopica, close-up of flower

Clinical Findings: A painful burning sensation of the lips and mouth results from ingestion. There is an inflammatory reaction, often with edema and blistering. Hoarseness, dysphonia, and dysphagia may result.

Management: The pain and edema recede slowly without therapy. Cool liquids or demulcents held in the mouth may bring some relief. Analgesics may be indicated. The insoluble oxalate in these plants does not cause systemic oxalate poisoning. Consultation with a Poison Control Center should be considered. See "Poisoning by Plants with Calcium Oxalate Crystals," p. 23.

Reference

Rauber A. Observations on the idioblasts of *Dieffenbachia*. J Toxicol Clin Toxicol 1985; 23:79–90.

Zephyranthes atamasco (L.) Herb.
(= *Amaryllis atamasco* L.)
Family: Amaryllidaceae

Common Names: Atamasco Lily, Fairy Lily, **Rain Lily**, Zephyr Lily

Description: The grassy leaves of *Zephyranthes* emerge from the ground and grow to a length of 1 foot. The single, erect flowers form on a hollow, leafless stalk and are usually white but may be tinged with purple. The plant is propagated by bulbs.

Distribution: *Zephyranthes* grows in wet areas from Virginia to Florida west to Alabama.

Toxic Part: The bulb is poisonous.

Toxin: Lycorine and related phenanthridine alkaloids (see *Narcissus*).

Clinical Findings: Ingestion of small amounts produces little or no symptoms. Large exposures may cause nausea, vomiting, abdominal cramping, diarrhea, dehydration, and electrolyte imbalance.

Zephyranthes atamasco

Management: Most exposures result in minimal or no toxicity. Intravenous hydration, antiemetics, and electrolyte replacement may be necessary for patients with severe gastrointestinal symptoms, particularly in children. Consultation with a Poison Control Center should be considered. See "Poisoning by Plants with Gastrointestinal Toxins," p. 28.

Reference

Jasperson-Schib R. Toxic Amaryllidaceae. Pharm Acta Helv 1970;45:424–433.

Zigadenus species (sometimes written *Zygadenus*)
Family: Liliaceae

Common Names: Alkali Grass, **Death Camas**, Hog's Potato, Poison Sego, Sand Corn, Soap Plant, Squirrel Food, Water Lily, Wild Onion

Description: These perennial herbs have grasslike leaves up to 1.5 feet long. The flowers form along the top of the central stalk and are usually yellow or whitish-green. Most species of *Zigadenus* have an onion-like bulb, but none has the characteristic onion odor.

Distribution: *Zigadenus* species are found throughout the United States except in the extreme southeast and Hawaii; they also are found across Canada and in Alaska.

Toxic Part: All parts of the plant, including the flowers, are toxic.

Zigadenus fremontii

Toxin: Zygadenine, zygacine, isogermidine, neogermidine, and protoveratridine, sodium channel activators.

Clinical Findings: Symptoms are predominantly neurological and cardiac. There is transient burning in the mouth after ingestion, followed after several hours by increased salivation, vomiting, diarrhea, and a tingling sensation in the skin. The patient may complain of headache, muscular weakness, and dimness of vision. Bradycardia and other cardiac dysrhythmias can be associated with severe blood pressure abnormalities. Coma may develop, and

convulsions may be a terminal event.

Management: Fluid replacement should be instituted with respiratory support if indicated. Heart rhythm and blood pressure should be monitored and treated with appropriate medications and supportive care. Recovery is generally complete within 24 hours. Consultation with a Poison Control Center should be strongly considered. See "Poisoning by Plants with Sodium Channel Activators," p. 32.

Zigadenus nuttallii

References

Grover J, Dahl B, Caravati M. Death camus: Mistaken identity at an herb farm. J Toxicol Clin Toxicol 1999;37:618–619.

Heilpern KL. *Zigadenus* poisoning. Ann Emerg Med 1995;25:259–262.

Peterson MC, Rasmussen GJ. Intoxication with foothill camas (*Zigadenus paniculatus*). J Toxicol Clin Toxicol 2003;41:63–65.

Wagstaff DJ, Case AA. Human poisoning by *Zigadenus*. J Toxicol Clin Toxicol 1987;25: 361–367.

Photographers' Credits

Adam, **Irina**

Acokanthera oppositifolia (2 photos), *Alocasia cv. Hilo's Beauty*, *Alocasia watsoniana*, *Alocasia × amazonica*, *Arum italicum*, *Aucuba japonica* (3 photos), *Brassaia actinophylla*, *Caladium bicolor* (2 photos), *Clivia nobilis* (2 photos), *Colocasia esculenta*, *Crassula argentea* (2 photos), *Cycas circinalis*, *Ephedra gerardiana* (2 photos), *Ficus benjamina*, *Gingko biloba*, *Hydrangea sp.*, *Ligustrum japonicum*, *Malus sp.*, *Prunus sp.*, *Ricinus communis*, *Spathiphyllum sp.*, *Swietenia mahagoni*, *Wisteria sinensis*, *Xanthosoma violacea*, *Zamia integrifolia*

Balick, **Michael**

Abrus precatorius (2 photos), *Adenium sp.*, *Adonis amurensis* (2 photos), *Allamanda cathartica* (2 photos), *Allium canadense* (2 photos), *Hippeastrum* 'Basuto', *Anemone coronaria*, *Pulsatilla vulgaris* 'Papageno', *Anthurium × roseum*, *Arum italicum*, *Atropa belladonna*, *Baptisia cv.* 'Purple Smoke', *Blighia sapida*, *Brassaia actinophylla*, *Calycanthus floridus cv.* 'Athens', *Catharanthus roseus* (3 photos), *Celastrus scandens*, *Cestrum nocturnum*, *Chelidonium majus*, *Chrysanthemum sp.* (3 photos), *Clematis jackmanii*, *Clematis cv.* 'Niobe,' *Colchicum autumnale* (3 photos), *Conium maculatum* (2 photos), *Convallaria majalis* (2 photos), *Crinum bulbispermum*, *Cryptostegia sp.* (2 photos), *Cycas revoluta*, *Datura stramonium* (2 photos), *Digitalis purpurea* (2 photos), *Epipremnum aureum*, *Euphorbia milii*, *Ficus benjamina*, *Ficus elastica*, *Galanthus nivalis*, *Hedera helix*, *Helleborus niger*, *Helleborus niger cv.* 'Maximus', *Hippobroma longiflora*, *Hura crepitans* & *Canavalia rusiosperma*, *Hydrangea macrophylla*, *Hypericum perforatum* (2 photos), *Iris pseudoacorus* (2 photos), *Jatropha curcas*, *Kalmia angustifolia*, *Kalmia latifolia* 'Pink Surprise', *Lantana camara* (2 photos), *Lonicerca tatarica*, *Manihot esculenta* (3 photos), *Melia azedarach*, *Monstera deliciosa*, *Narcissus poeticus*, *Narcissus pseudonarcissus*, *Narcissus sp.*, *Nerium oleander* (2 photos), *Ornithogalum thyrsoides* (2 photos), *Ornithogalum umbellatum*, *Pentalinon luteum*, *Philodendron selloum*, *Pieris floribunda*, *Pieris japonica* (2 photos), *Prunus pendula var. adscendens* (2 photos), *Prunus serotina*, *Prunus laurocerasus*, *Pteridium acquilinium*, *Rhamnus cathartica*, *Rhododendron cv.* 'Yaku Princess', *Rhododendron cv.* 'Rosebud', *Rhododendron × PMJ*, *Robinia pseudoacacia*, *Scilla peruviana*, *Scilla siberica*, *Sophora secundiflora*, *Symplocarpus foetidus*, *Toxicodendron radicans*, *Veratrum viride*, *Wisteria sinensis*.

Cheatham, **Scooter**

Astralagus mollissimus, *Astralagus wootonii*, *Karwiskia humboldtiana* (2 photos), *Sophora secundiflora* (2 photos), *Symphoricarpos orbiculatus*

Foster, **Steven**

Aconitum napellus, *Actaea rubra*, *Aesculus californica*, *Aesculus glabra*, *Aesculus hippocastanum* (2 photos), *Aloe vera*, *Anemone canadensis*, *Pulsatilla patens*, *Pulsatilla vulgaris*, *Arisaema dracontium*, *Arisaema triphyllum*, *Atropa belladonna*, *Baptisia alba*, *Blighia sapida*, *Capsicum annuum*, *Caulophyllum thalictroides* (2 photos), *Chelidonium majus*, *Cicuta maculata*, *Clematis virginiana*, *Clivia miniata*, *Colocasia esculenta*, *Conium maculatum*, *Crinum asiaticum*, *Datura wrightii*, *Digitalis purpurea*, *Dirca palustris*, *Eriobotrya japonica*, *Euonymus americanus*, *Euonymus atropurpurea*, *Euphorbia marginata*, *Euphorbia pulcherrima*, *Gelsemium sempervirens*, *Gymnocladus dioicus*, *Hydrastis canadensis* (3 photos), *Hypericum perforatum*, *Ilex opaca*, *Lantana camara*, *Ligustrum lucidum* (2 photos), *Lobelia cardinalis*, *Lobelia inflata*, *Lobelia siphi-*

litica, Lupinus perennis, Lycoris radiata, Lyonia sp., Menispermum canadense, Momordica charantia, Nerium oleander, Nicotiana rustica, Nicotiana tabacum, Phoradendron serotinum (2 photos), *Physalis crassifolia, Phytolacca americana, Podophyllum peltatum* (2 photos), *Poncirus trifoliata, Ranunculus acris, Ranunculus sceleratus, Rhamnus cathartica, Rheum × cultorum, Ricinus communis* (2 photos), *Robinia pseudoacacia, Sambucus nigra, Sambucus nigra* subsp. *caerulea, Sambucus nigra* subsp. *canadensis, Sanguinaria canadensis, Schinus molle, Schinus terebinthifolius, Senecio douglasii, Senecio jacobea, Spigelia marilandica, Symphoricarpos albus, Taxus brevifolia* (2 photos), *Thevetia peruviana, Veratrum californicum, Viscum album, Zantedeschia aethiopica, Zigadenus fremontii, Zigadenus nuttallii*

Goltra, Peter
Calophyllum inophyllum (2 photos)

Gromping, Hans-Wilhelm
Hyoscyamus niger

Henderson, Flor
Caryota mitis (embryo)

Henderson, Andrew
Caryota urens, Caryota gigas

Howard, Richard A.
Aconitum napellus, Actaea pachypoda, Actaea rubra, Actaea spicata, Adenium sp., Aleurites moluccana, Aloe speciosa, Hippeastrum sp., Hippeastrum puniceum, Anthurium × ferrierense, Anthurium wildenowii, Baptisia tinctoria, Caesalpinia bonduc (2 photos), *Caesalpinia gilliesii, Calophyllum inophyllum, Calotropis gigantea, Calotropis procera* (2 photos), *Caltha palustris, Calycanthus floridus* (2 photos), *Cassia fistula* (3 photos), *Celastrus scandens, Celastrus scandens* & *C. orbiculatus, Cestrum nocturnum, Clematis paniculata, Clusia rosea* (3 photos), *Daphne mezereum, Datura metel, Datura metel cv.* Cornucopaea, *Brugmansia candida, Datura sanguinea, Dieffenbachia seguine, Dirca palustris, Euonymus americanus, Euonymus europaeus* (2 photos), *Euphorbia cyathophora, Euphorbia lactea* (3 photos), *Euphorbia lathyris, Euphorbia milii var. splendens, Euphorbia tirucalli, Gelsemium sempervirens, Gloriosa superba, Hedera helix, Heliotropium indicum, Hippomane mancinella* (2 photos), *Hura crepitans, Hymenocallis caribaea, Hymenocallis declinata, Ilex opaca, Ilex vomitoria, Iris germanica, Jatropha curcas, Jatropha podagrica* (2 photos), *Kalmia latifolia* (2 photos), *Laburnum anagyroides, Leucaena leucocephala* (2 photos), *Leucothoe sp., Lonicera periclymenum, Lycium carolinianum, Momordica charantia Monstera deliciosa* (2 photos), *Nicotiana glauca, Pedilanthus tithymaloides* (2 photos), *Pentalinon luteum, Philodendron sp., Physalis alkekengi* (2 photos), *Poncirus trifoliata, Ricinus communis, Rivina humilis, Sambucus racemosa, Sapindus saponaria, Schinus molle, Scilla hispanica, Sesbania grandifolia* (2 photos), *Solandra guttata* (2 photos), *Solanum dulcamara* (2 photos), *Solanum mammosum, Solanum seaforthianum, Spathiphyllum sp., Spigelia anthelmia, Strychnos nux-vomica, Taxus cuspidata cv. 'Aurescens', Urginea maritimae, Wisteria floribunda, Xanthosoma sagittafolium.*

Kronenberg, Fredi
Sophora secundiflora

Lighty, Richard W.
Abrus precatorius, Adonis annua, Arum italicum, Caesalpinia pulcherrima, Caryota sp., Cryptostegia grandiflora, Cryptostegia madagascariensis, Daphne mezereum, Duranta repens (2

photos), *Hydrangea macrophylla, Laburnum anagyroides, Lonicera periclymenum, Lonicera tatarica, Manihot esculenta, Rhodotypos scandens* (2 photos), *Scilla sinensis, Zephyranthes atamasco*

Linney, George K.
Corynocarpus laevigatus

Mickel, John
Pteridium aquilinum

Nee, Michael
Cestrum diurnum, Brugmansia suaveolens, Solandra grandiflora, Solanum americanum (2 photos), *Solanum mammosum, Solanum pseudocapsicum, Solanum tuberosum*

The New York Botanical Garden
Illustrations from LuEsther T. Mertz Library:
Aethusa cynapium, Caltha palustris, Cassia occidentalis, Coriaria myrtifolia (2 illustrations), *Corynocarpus laevigatus, Daphne mezereum, Echium vulgare, Gymnocladus dioicus, Melia azedarach, Menispermum canadense, Myoporum laetum, Oenanthe crocata, Pachyrhizus erosus, Swietenia mahagoni, Toxicodendron diversilobum, Toxicodendron radicans, Toxicodendron vernix*
Photos from Archives:
Alocasia macrorrhiza cv. 'variegata', *Crotalaria sagittalis, Pernettya mucronata, Schoenocaulon drummondii*

Nixon, Kevin
Pernettya mucronata

Schoepke, Thomas
Ligustrum vulgare, Myoporum laetum

Stevenson, Dennis Wm.
Zamia pumila (2 photos)

Index

Note: page numbers followed by f indicate figures.
Note: page numbers followed by t indicate tables.

Disclaimer

This book is a work of reference and is not intended to supply medical advice to any particular individual. Readers should always consult their personal physicians for medical advice and treatment. The authors, editors, and publisher of this work have checked with sources believed to be reliable in their efforts to confirm the accuracy and completeness of the information presented herein and that the information is in accordance with the standard practices accepted at the time of publication. However, neither the authors, editors, and publisher, nor any party involved in the creation and publication of this work warrant that the information is in every respect accurate and complete, and they are not responsible for errors or omissions or for any consequences from the application of the information in this book. In light of ongoing research and changes in clinical experience and in governmental regulations, readers are encouraged to confirm the information contained herein with additional sources, in particular as it pertains to drug dosage and usage. Readers are urged to check the package insert for each drug they plan to administer for any change in indications or dosage or for additional warnings or precautions, especially for new or infrequently used drugs. This book does not purport to be a complete presentation of all poisonous plants, and the genera, species, and cultivars discussed or pictured herein are but a small fraction of the plants known to have caused poisonings to humans and animals that might be found in the wild, in an urban or suburban landscape, or in a home. Given the global movement of plants, we would expect continual introduction of species having toxic properties to the regions discussed in this book. We have made every attempt to be botanically accurate, but regional variations in plant names, growing conditions, and availability may affect the accuracy of the information provided. A positive identification of an individual plant is most likely when a freshly collected part of the plant containing leaves and flowers or fruits is presented to a knowledgeable botanist or horticulturist. Poison Control Centers generally have relationships with the botanical community should the need for plant identification arise. Medicine is an everchanging science and subject to interpretation. We have attempted to provide accurate medical descriptions of clinical management strategies, but there is no substitute for direct interaction with an expert clinician for assisting in patient care as well as a trained botanist or horticulturist for plant identification. **In cases of exposure or ingestion, contact a Poison Control Center (1-800-222-1222), a medical toxicologist, another appropriate healthcare provider, or an appropriate reference resource.**